Memoirs
of the
American Mathematical Society

Volume 226 • Number 1061 (second of 5 numbers) • November 2013

Isolated Involutions in Finite Groups

Rebecca Waldecker

ISSN 0065-9266 (print)   ISSN 1947-6221 (online)

American Mathematical Society
Providence, Rhode Island

Library of Congress Cataloging-in-Publication Data

Waldecker, Rebecca, 1979-
  Isolated involutions in finite groups / Rebecca Waldecker.
    pages cm. – (Memoirs of the American Mathematical Society, ISSN 0065-9266 ; volume 226, number 1061)
  "November 2013, volume 226, number 1061 (second of 5 numbers)."
  Includes bibliographical references and index.
  ISBN 978-0-8218-8803-2 (alk. paper)
    1. Involutes (Mathematics). 2. Finite groups.. 3. Solvable groups.. 4. Feit-Thompson theorem.. 5. Glauberman, G., 1941- . I. Title.
  QA557.W35   2013
  512′.23–dc23   2013025509

DOI: http://dx.doi.org/10.1090/S0065-9266-2013-00684-3

Memoirs of the American Mathematical Society

This journal is devoted entirely to research in pure and applied mathematics.

Publisher Item Identifier. The Publisher Item Identifier (PII) appears as a footnote on the Abstract page of each article. This alphanumeric string of characters uniquely identifies each article and can be used for future cataloguing, searching, and electronic retrieval.

Subscription information. Beginning with the January 2010 issue, *Memoirs* is accessible from www.ams.org/journals. The 2013 subscription begins with volume 221 and consists of six mailings, each containing one or more numbers. Subscription prices are as follows: for paper delivery, US$795 list, US$636 institutional member; for electronic delivery, US$700 list, US$560 institutional member. Upon request, subscribers to paper delivery of this journal are also entitled to receive electronic delivery. If ordering the paper version, add US$10 for delivery within the United States; US$69 for outside the United States. Subscription renewals are subject to late fees. See www.ams.org/help-faq for more journal subscription information. Each number may be ordered separately; *please specify number* when ordering an individual number.

Back number information. For back issues see www.ams.org/bookstore.

Subscriptions and orders should be addressed to the American Mathematical Society, P. O. Box 845904, Boston, MA 02284-5904 USA. *All orders must be accompanied by payment.* Other correspondence should be addressed to 201 Charles Street, Providence, RI 02904-2294 USA.

Copying and reprinting. Individual readers of this publication, and nonprofit libraries acting for them, are permitted to make fair use of the material, such as to copy a chapter for use in teaching or research. Permission is granted to quote brief passages from this publication in reviews, provided the customary acknowledgment of the source is given.

Republication, systematic copying, or multiple reproduction of any material in this publication is permitted only under license from the American Mathematical Society. Requests for such permission should be addressed to the Acquisitions Department, American Mathematical Society, 201 Charles Street, Providence, Rhode Island 02904-2294 USA. Requests can also be made by e-mail to reprint-permission@ams.org.

Memoirs of the American Mathematical Society (ISSN 0065-9266 (print); 1947-6221 (online)) is published bimonthly (each volume consisting usually of more than one number) by the American Mathematical Society at 201 Charles Street, Providence, RI 02904-2294 USA. Periodicals postage paid at Providence, RI. Postmaster: Send address changes to Memoirs, American Mathematical Society, 201 Charles Street, Providence, RI 02904-2294 USA.

© 2013 by the American Mathematical Society. All rights reserved.
Copyright of individual articles may revert to the public domain 28 years
after publication. Contact the AMS for copyright status of individual articles.
This publication is indexed in *Mathematical Reviews*®, *Zentralblatt MATH*, *Science Citation Index*®, *Science Citation Index*^{TM}*-Expanded*, *ISI Alerting Services*^{SM}, *SciSearch*®, *Research Alert*®, *CompuMath Citation Index*®, *Current Contents*®/*Physical, Chemical & Earth Sciences*. This publication is archived in
Portico and *CLOCKSS*.
Printed in the United States of America.

∞ The paper used in this book is acid-free and falls within the guidelines
established to ensure permanence and durability.
Visit the AMS home page at http://www.ams.org/

10 9 8 7 6 5 4 3 2 1   18 17 16 15 14 13

To Lars

Contents

Chapter 1.	Introduction	1
Chapter 2.	Preliminaries	7
2.1.	Definitions and Notation	7
2.2.	General Results	9
2.3.	A Nilpotent Action Result	14
Chapter 3.	Isolated Involutions	17
Chapter 4.	A Minimal Counter-Example to Glauberman's Z*-Theorem	23
Chapter 5.	Balance and Signalizer Functors	29
Chapter 6.	Preparatory Results for the Local Analysis	37
6.1.	The Bender Method	37
6.2.	t-Minimal Subgroups, Pushing Down and Uniqueness Results	40
Chapter 7.	Maximal Subgroups Containing C	45
Chapter 8.	The 2-rank of $O_{2',2}(C)$	55
8.1.	Involutions in $O_{2',2}(C)\setminus\{z\}$	55
8.2.	The Proof of Theorem **B**	63
Chapter 9.	Components of $\overline{C}$ and the Soluble Z*-Theorem	69
Chapter 10.	Unbalanced Components	75
Chapter 11.	The 2-Rank of G	79
Chapter 12.	The F*-Structure Theorem	87
Chapter 13.	More Involutions	93
13.1.	Preliminary Results	93
13.2.	The Symmetric Case	102
13.3.	The General Case	111
Chapter 14.	The Endgame	123
Chapter 15.	The Final Contradiction and the Z*-Theorem for $\mathcal{K}_2$-Groups	145
Bibliography		147
Index		149

Abstract

This text provides a new proof of Glauberman's Z*-Theorem under the additional hypothesis that the simple groups involved in the centraliser of an isolated involution are known simple groups.

Received by the editor November 19, 2011, and, in revised form, February 15, 2012.
Article electronically published on March 15, 2013; S 0065-9266(2013)00684-3.
2010 *Mathematics Subject Classification.* Primary 20E25, 20E34.
Key words and phrases. Finite group, Z*-Theorem, isolated involution.
This work was partially supported by the Studienstiftung des deutschen Volkes and by the Leverhulme Trust.
Affiliation at time of publication: Martin-Luther-Universität Halle-Wittenberg, Institut für Mathematik, 06099 Halle (Saale), Germany; email: rebecca.waldecker@mathematik.uni-halle.de.

©2013 American Mathematical Society

CHAPTER 1

Introduction

The protagonist of this text is one of the main results in Glauberman's article "Central Elements in Core-Free Groups" from the year 1966:

GLAUBERMAN'S Z*-THEOREM. *Suppose that G is a finite group and that $z \in G$ is an isolated involution. Then $\langle z \rangle O(G) \trianglelefteq G$.*

Some explanation is needed here: We say that an element z in a finite group G is an **involution** if it has order 2. Then z is called **isolated in** G if and only if the only conjugate of z in G commuting with z is z itself. Another way of expressing this that can be found in the literature (for example in [**Gor82**]) is to specify a Sylow 2-subgroup S of G containing z and to say that z is **isolated in S with respect to** G if and only if z itself is the only conjugate of z in S. Moreover, the term $O(G)$ is the standard abbreviation for the subgroup $O_{2'}(G)$, i.e. the largest normal subgroup of G of odd order. This subgroup is sometimes referred to as the **core of** G. Roughly speaking, the Z*-Theorem says that isolated involutions are central modulo the core. If we denote by $Z^*(G)$ the full pre-image in G of the factor group $Z(G/O(G))$, then the Z*-Theorem can be re-phrased in the following way:

Every isolated involution of a finite group G is contained in $Z^*(G)$.

The reader will find versions of the Z*-Theorem with a variety of different notation and emphasis in the literature. In Glauberman's original article, the result that is closest to the version stated here is Theorem 1. Glauberman explains that his Z*-Theorem "originated as a conjecture in loop theory". A special case of this conjecture had been proved earlier by Fischer (see [**Fis64**] and additional comments in Chapter 3). If a finite group G has cyclic or quaternion Sylow 2-subgroups, then the unique involution z in a Sylow 2-subgroup of G is isolated in G. Before Glauberman's theorem it was already known, because of results by Burnside and by Brauer and Suzuki, that $\langle z \rangle O(G) \trianglelefteq G$ in these cases. Therefore the Z*-Theorem can be viewed as a generalisation in particular of the Brauer-Suzuki result, a viewpoint taken for example by Gorenstein in [**Gor82**].

As it turned out (and as Gorenstein emphasises in [**Gor82**]), Glauberman's Z*-Theorem became one of the most fundamental local group theoretic results in the context of the Classification of Finite Simple Groups. To illustrate this, let us suppose that G is a non-abelian finite simple group. Then the Odd Order Theorem of Feit and Thompson (see [**FT63**]) says that G has even order and so it follows that G contains an element t of order 2. If t is isolated in G, then the Z*-Theorem forces $t \in Z^*(G)$. But G is simple and has even order, so $O(G) = 1$ and hence $t \in Z(G)$. This is impossible because G is simple and not abelian. We conclude that t cannot be isolated in G. In fact this can be phrased as a special consequence of the Z*-Theorem:

A non-abelian finite simple group does not contain any isolated involutions.

This can be used as a starting point for understanding the 2-structure of a finite group and therefore it is not surprising that, in connection with classification results based on types of Sylow 2-subgroups or involution centralisers, the Z*-Theorem is a powerful tool. Another important consequence is Glauberman's main result from [**Gla66b**] that leads to proofs of special cases of the Schreier Conjecture. (This conjecture says that the outer automorphism group of a finite simple group is soluble.) It is worth emphasising Glauberman's progress in this direction because, still, no classification-free proof of the full Schreier Conjecture is known.

Returning to the Z*-Theorem, its discussion in later literature usually not only points out its consequences for the, at the time, ongoing effort towards the Classification of Finite Simple Groups, but also the fact that the proof uses elegant arguments from block theory. Its proof is actually often found as an illustration of the power of Brauer's Main Theorems. Also, thinking in the direction of future generalisations, why not extend the notion of an isolated involution to an "isolated element of prime order p" and attempt to find an "odd version" of the Z*-Theorem? This leads to two natural questions:

(1) How difficult is it (if at all possible) to prove the Z*-Theorem with local group theoretic methods?
(2) What could be a reasonable conjecture that generalises the Z*-Theorem for odd primes? Is it possible to prove such a conjecture?

The reasons for Question (1) are, from our point of view, both philosophical and practical. Philosophically speaking, it might be more satisfying if a result that plays such an important role in local group theory could be understood from a local perspective, giving also some indication of the strength of local techniques. A more practical viewpoint comes in as soon as the difficulty of such a task becomes apparent. Even if finding a new proof fails, it can be expected that interesting results will emerge and that a number of group theoretic arguments will be refined and extended on the way.

In this text we prove that the Z*-Theorem holds for all groups where, roughly speaking, the simple sections in an involution centraliser are known simple groups. Although some of the background results require representation theory (as for example [**FT63**] or [**Gla74**]), the proof itself is based on local group theoretic methods and thus we give an almost complete answer to Question (1) above. We leave it to the reader to decide how difficult this new proof is – it is certainly much longer and more technical than Glauberman's original proof and, maybe unsurprisingly, it involves quite a few different techniques that play a role in the Classification of Finite Simple Groups.

Concerning Question (2), it is known by the Classification that the Z*-Theorem generalises for odd primes in a natural way. But seeing this requires the use of the Classification in its full strength (see for example [**GR93**]) and does, so far, not give much insight into why such a result holds. Special cases have been proved for example by Rowley (for the prime 3, see [**Row81**]) and by Broué (see [**Bro83**]). There is an ongoing effort from group theorists and from representation theorists to make some progress towards proving an "Odd Z*-Theorem" without using the full and immediate strength of the Classification. It is our hope that the local approach to Glauberman's original result will shed some light on what group theoretic tools might have a role to play.

The remainder of this introduction gives an overview of the strategy and some indication of what happens mathematically in which part of this text. All groups mentioned here are finite. In earlier work (see [**Wal09**]) we show that if G is a minimal counter-example to Glauberman's Z*-Theorem and if C is the centraliser of an isolated involution $z \in G$ with $z \notin Z^*(G)$, then $C/O(C)$ possesses at least one component. In particular C is not soluble. This work plays a role in our general approach here and is therefore partly included.

Chapter 2 starts with some preparation: setting up notation, recalling or specifying definitions and stating background results. In order to make this text fairly self-contained, we give precise references or proofs for all results listed there. Then, in Chapter 3, we turn to groups with isolated involutions. We establish a crucial result (Theorem 3.6) that implies, for example, that a minimal counter-example to the Z*-Theorem is generated by two involution centralisers and this is the basis for a counting argument at the end of Chapter 8. It is of similar importance that for every isolated involution in a group G and for every prime p there exists a Sylow p-subgroup of G that is normalised by this involution. These results suggest that isolated involutions behave as if acting coprimely on every subgroup that they normalise. The content of Chapter 3 has mostly appeared before in [**Wal09**].

Then it is time to look at a minimal counter-example to Glauberman's Theorem. In Chapter 4 we set up our first working hypothesis which says, in a nutshell, that G is a group with an isolated involution z such that G provides a minimal counter-example to the Z*-Theorem. In particular $z \notin Z^*(G)$, but the Z*-Theorem holds in every proper subgroup of G and in proper factor groups. We set $C := C_G(z)$ and assume this hypothesis in the remainder of the discussion, in particular in Theorems **A**, **B**, **C** and **D**. We prove initial consequences of this setup, again following the exposition in [**Wal09**] in many places. For example it turns out that G is almost simple and that every maximal subgroup of G containing an isolated involution is primitive (as defined on page 7). Here we should point out that the p Complement Theorem and the Brauer-Suzuki Theorem are used to show that G does not have cyclic or quaternion Sylow 2-subgroups.

Then we exploit more specific properties of G and z in Chapter 5. We introduce a variety of balance notions and the concept of signalizer functors. Then we use the fact that our balance conditions usually fail in G and study the consequences for the structure of G and specifically of C. Some arguments in this chapter are inspired by Goldschmidt's work in [**Gol72**] and [**Gol75**], for example we present and apply signalizer functors that he uses in these papers.

After some preparation in Chapter 6, we work towards the proof of our first important result in Chapter 7.

THEOREM **A**. *Suppose that M is a maximal subgroup of G containing C. If possible, choose M such that there exists a prime q with $O_q(M) \neq 1 = C_{O_q(M)}(z)$. Then one of the following holds:*
- *$M = C$.*
- *There exists an odd prime p such that $F^*(M) = O_p(M)$.*
- *$E(M) \neq 1$.*

One of the most important ingredients for the proof of Theorem **A** and also for arguments in later sections is the so-called Bender Method. It is therefore introduced at the beginning of Chapter 6. In particular, Lemma 6.2 and the Infection

Theorem 6.3 are mainly an adaptation of results of Bender's for our situation to simplify quotations. Then we prove some preparatory results about isolated involutions in proper subgroups of G that are also needed for later sections. Finally we set up our working hypothesis at the beginning of Chapter 7, assuming that M is a maximal subgroup of G that properly contains C and that is chosen in a technically suitable way. We show that, if M is not of characteristic p, then for certain subgroups X of $F(M)$, we can force $N_G(X)$ to be contained in a unique maximal subgroup of G, namely in M. Then we assume that M is in fact a counter-example to Theorem **A** and we find a prime p such that $O_p(M)$ is cyclic and z inverts it (Lemma 7.5). This leads to a non-trivial normal subgroup of G that is contained in M, giving a contradiction. Later it will emerge that Theorem **A** can sometimes be strengthened or that similar statements hold for other involution centralisers (see for example Theorem 13.23).

In the remainder of the text the strategy is to determine the structure of $\overline{C} := C/O(C)$ as far as possible and then to analyse several involution centralisers at the same time. First we consider the situation where $O_{2',2}(C)$, the full pre-image of $O_2(\overline{C})$, possesses an elementary abelian subgroup of order 4. This case is excluded in Chapter 8. The path that we follow is similar to that in [**Wal09**] at first, but then some new arguments are necessary. So we conclude:

THEOREM **B**. *z is the unique involution in $O_{2',2}(C)$.*

Next we turn to the components of $\overline{C}$. In Chapter 9 we restrict their number and shape, still not using any additional hypothesis.

THEOREM **C**. *$\overline{C}$ possesses at least one and at most three components. If there are three components, then they are all of type A_7 or $PSL_2(q)$ for some odd number $q \geq 5$.*

Here a component $\overline{E}$ of $\overline{C}$ is said to be **of type** A_n (or $PSL_2(q)$) if $\overline{E}/Z(\overline{E})$ is isomorphic to A_n (or to $PSL_2(q)$). In order to prove Theorem **C**, we apply results from Section 5, in particular Goldschmidt's notion of core-separated subgroups and signalizer functor theory. In the background, the Gorenstein-Walter-Theorem on groups with dihedral Sylow 2-subgroups plays a role. It is here that the Soluble Z*-Theorem follows and we give the arguments, for completeness, although they are explained in [**Wal09**] as well. (This is actually the last place where we re-state results from [**Wal09**].) We do not need the full strength of Theorem **C** for the Soluble Z*-Theorem, but only the fact that $\overline{C}$ has components at all.

THE SOLUBLE Z^*-THEOREM.
Suppose that G is a finite group and that $z \in G$ is an isolated involution. If $C_G(z)$ is soluble, then $\langle z \rangle O(G) \trianglelefteq G$.

Beginning in Chapter 10, and in all later sections, we suppose that whenever $\overline{E}$ is a component of $\overline{C}$, then the simple group $\overline{E}/Z(\overline{E})$ is known. In particular we assume this hypothesis in Theorem **D**. We begin to understand the structure of $F^*(\overline{C})$ by looking at the case where $r_2(G) \geq 4$. In order to obtain information about the possible types of components, we prove that $\overline{C}$ has so-called "unbalanced" components, using results from Chapter 5. Although obtained differently, the functor that we apply for these arguments is the same as in Proposition 4.65

of [**Gor82**]. Our additional hypothesis about components of $\overline{C}$ comes in when we argue with failure of 2-balance and quote a result in [**Gor82**]. The next step is then to exclude components of type A_n where n is at least 10. This is done in Lemma 10.5, again with a signalizer functor, and enables us to bound the 2-rank of G in Section 11.

THEOREM **D**. *G has 2-rank 2 or 3.*

The main argument for the proof of this result is to assume that $r_2(G) \geq 4$ and to show that, as a consequence, conditions about balance or core-separated subgroups in C are violated. The technical details that we encounter are mainly dealt with in a series of lemmas, excluding particular configurations for $F^*(\overline{C})$ one by one, whilst the method is usually to either construct a signalizer functor (and reach a contradiction) or to analyse a "failure of balance" situation. After the 2-rank of G is restricted, we can describe the structure of $F^*(\overline{C})$ in much detail. This information is collected in the F*-Structure Theorem 12.6, subdivided into four lists referred to as List I, II, III and IV. Once it is established, we apply the F*-Structure Theorem by saying that "$F^*(\overline{C})$ is as on List I (or II, III or IV)".

Based on this information, we go back to analysing maximal subgroups containing the centraliser of an involution in Chapter 13. We choose our involutions carefully and reveal enough of the structure of their centralisers to bring the Bender Method into the picture again. Having control over centralisers of (at least some) involutions in G, the stage is set for the "endgame", a final situation that needs to be analysed. We attack this in Chapter 14 and, based on this, derive a final contradiction in the last section. This is also where an independent version of the Z*-Theorem is stated, with an explanation of why our work proves this version. For this purpose, we define a $\mathcal{K}_2$-group to be a group X where for every isolated involution $t \subset X$ and every subgroup H of X containing t, the simple groups involved in $C_H(t)/O(C_H(t))$ are known simple groups. At the end of Chapter 15, we prove:

THE Z*-THEOREM FOR $\mathcal{K}_2$-GROUPS. *Suppose that G is a $\mathcal{K}_2$-group and that $z \in G$ is an isolated involution. Then $\langle z \rangle O(G) \trianglelefteq G$.*

In particular, whenever a result about finite groups is proved under the hypothesis that every proper simple section is known, then this weaker version of the Z*-Theorem can be applied. This is relevant, for example, to a minimal counterexample to the Classification Theorem itself, but also to many results in progress that contribute to new, different strategies for the Classification.

Acknowledgements

It is a pleasure to sincerely thank Helmut Bender with whom the Z*-Project started in 2003. The author is also indebted to a number of colleagues for their encouragement and helpful suggestions at different stages of this work, especially to Luke Morgan, Chris Parker, Steve Smith and Ron Solomon. The thorough reading of the manuscript by the referee is much appreciated and the comments lead to substantial improvements of this text.

CHAPTER 2

Preliminaries

In this section we introduce the notation that is used in this text and we state general results that are applied so that they can be quoted explicitly when needed. Most of these results are fairly standard and can be found, for example, in group theory books – in these cases, we give a reference. Otherwise we give a proof. All groups are meant to be finite and we follow the notation in standard group theory books such as [**Asc00**] and [**KS04**]. We also use throughout, without further reference, that groups of odd order are soluble ([**FT63**]). In this section let X be a group, let π be a set of primes and let p and q be prime numbers.

2.1. Definitions and Notation

– For all $n \in \mathbb{N}$, we denote by n_p the largest power of p dividing n.

– For all $x \in X$, all subgroups $Y \leq X$ and all subsets $U \subseteq X$ we define $x^U := \{x^u \mid u \in U\}$ and $Y^U := \{Y^u \mid u \in U\}$.

– By $H \max X$ we mean that H is a maximal subgroup of X.

– A subgroup H of X is said to be **primitive** if and only if, for all $1 \neq U \trianglelefteq H$, we have that $N_X(U) = H$. A typical example for a primitive group is a maximal subgroup of a simple group.

– An involution $t \in X$ is **isolated in** X if and only if $C_X(t) \cap t^X = \{t\}$.

– If an involution t acts on a subset Y of X, then $I_Y(t)$ denotes the set of elements of Y that are inverted by t.

– The largest normal π'-subgroup of X is usually denoted by $O_{\pi'}(X)$. Then $Z_\pi^*(X)$ denotes the full pre-image of $Z(X/O_{\pi'}(X))$ in X.

– As a special case of the above, the largest normal subgroup of odd order of X is abbreviated as $O(X)$ (sometimes referred to as the **core** of X in the literature). Then $Z^*(X)$ denotes the full pre-image of $Z(X/O(X))$ in X (and is hence an abbreviation of $Z_{\{2\}}^*(X)$).

– To simplify notation, we set $F_\pi(X) := O_\pi(F(X))$.

– If X is a p-group, then $ZJ(X)$ denotes the centre of the Thompson subgroup of X, see for example on page 162 in [**Asc00**].

– X is **quasi-simple** if and only if $X \neq 1$, X is perfect (i.e. $X' = X$) and $X/Z(X)$ is simple.

– $O^p(X)$ denotes the smallest normal subgroup of X that has a p-factor group. We say that X is p-**perfect** if and only if $X = O^p(X)$. We denote by $O^\infty(X)$ the smallest normal subgroup of X that has a soluble factor group.

– X is **of characteristic** p if and only if $F^*(X) = O_p(X)$. We denote this by $\operatorname{char}(X) = p$. If the prime is supposed to be unspecified, then we just say that X has

prime characteristic. A special case is that X has **odd prime characteristic**, meaning that there exists an odd prime p such that $F^*(X) = O_p(X)$.

– We say that a subgroup A of X is **centraliser closed** if and only if $C_X(A) \leq A$. If A is also abelian, then this implies that $A = C_X(A)$ whence we also say that A is **self-centralising**. For example, if X has characteristic p, then $O_p(X)$ is centraliser closed but not necessarily self-centralising.

– Let $A, B \leq X$ be subgroups such that AB is a subgroup of X. We say that AB is a **central product** and write $A * B$ if and only if $[A, B] = 1$.

– Let $A, B \leq X$ be 2-subgroups such that $[A, B] = 1$. If $a \in A$ and $b \in B$ are elements of order 4 such that $a^2 = b^2$, then ab is an involution and we say that this involution is **diagonal** in $A * B$.

– A **component** of X is a quasi-simple subnormal subgroup of X. We denote the (central) product of all components of X by $E(X)$.

– Let L be a simple group. A component E of X is said to be **of type** L if and only if $E/Z(E) \simeq L$.

– $O_{\pi',F^*}(X)$ denotes the preimage of $F^*(X/O_{\pi'}(X))$ in X. The subgroups $O_{\pi',\pi}(X)$, $O_{\pi',F}(X)$ and $O_{\pi',E}(X)$ are defined similarly.

We simplify notation if π consists of a single prime, for example we write $O_{2',2}(X)$ instead of $O_{2',\{2\}}(X)$.

– A subnormal subgroup E of X is a π-**component** of X if and only if $E = O^\infty(E)$ and $E/O_{\pi'}(E)$ is quasi-simple. The set of all π-components of X is denoted by $\mathcal{L}_\pi(X)$. The most important special case for us occurs for the prime 2 where we write $\mathcal{L}_2(X)$ for the set of 2-components (i.e. $\{2\}$-components) of X.

– A convenient abbreviation is $L(X) := O^{2'}(O_{2',E}(X))$.

– By $r_p(X)$ we denote the p-**rank of** X. This means that if $n \in \mathbb{N}$ is the largest number such that X possesses an elementary-abelian p-subgroup of order p^n, then we set $r_p(X) := n$. If there is no ambiguity about the prime we are referring to (for example because X is a p-group), then we only write $r(X)$ for the rank.

– Let $P \in \mathrm{Syl}_p(X)$. Then we say that X is p-**nilpotent** if and only if $X = O_{p'}(X)P$. (Another way of expressing this that can be found in the literature is that "X **has a normal** p-**complement**." .)

– Suppose that A is a group acting on X. The action of A on X is **nilpotent** if and only if there exists a $k \in \mathbb{N}$ such that $\underbrace{[...[[X, A], A]..., A]}_{k} = 1$.

– Suppose that $A \leq X$. Then by $\mathcal{U}_X(A, \pi)$ we denote the set of A-**invariant** π-**subgroups of** X. We write $\mathcal{U}_X^*(A, \pi)$ for the set of maximal members of $\mathcal{U}_X(A, \pi)$ with respect to inclusion.

– If $n \in \mathbb{N}$, then by $X \simeq C_n$ we mean that X is cyclic of order n.

– We say that X is **quaternion** if and only if $X \simeq Q_{2^n}$ for some $n \geq 3$ (rather then saying "generalised quaternion").

– For all $n \in \mathbb{N}$, we denote the symmetric (alternating) group of degree n by $\mathcal{S}_n$ (A_n). We write $2A_n$ for the quasi-simple group that has a centre of order 2 and that modulo its centre is isomorphic to A_n (non-split). Similarly the notation $3A_7$, $3PSL_2(9)$ and $2J_2$ is used.

2.2. General Results

LEMMA 2.1. *Suppose that a π-group P acts on a π'-group Q.*

(1) *If N is a P-invariant normal subgroup of Q, then $C_{Q/N}(P) = C_Q(P)N/N$.*

(2) *$Q = [Q,P]C_Q(P)$ and $[Q,P] = [Q,P,P]$. If Q is abelian, then $Q = [Q,P] \times C_Q(P)$.*

(3) *If Q is the product of two P-invariant subgroups Q_1 and Q_2, then $C_Q(P) = C_{Q_1}(P)C_{Q_2}(P)$.*

(4) *If P is an elementary-abelian, non-cyclic p-group, then*
$$Q = \langle C_Q(A) \mid A \leq P,\ |P:A| = p \rangle \text{ and}$$
$$[Q,P] = \langle [C_Q(A), P] \mid A \leq P,\ |P:A| = p \rangle.$$
If P has order 4, e.g. $P = \{1, x, y, xy\}$, and if Q is nilpotent, then $Q = C_Q(x)C_Q(y)C_Q(xy)$. Hence if $C_Q(x) \leq C_Q(y)$, then $I_Q(y) \subseteq I_Q(x)$.

(5) *If Q is a q-group for some odd prime $q \in \pi'$ and if P centralises every element of order q in Q, then $[Q,P] = 1$.*

(6) *If Q is nilpotent and P centralises a centraliser closed subgroup of Q, then P centralises Q.*

(7) *Let $r \in \pi'$. Then $\mathcal{M}_Q^*(P,r) \subseteq \mathrm{Syl}_r(Q)$ and $C_{QP}(P)$ is transitive on $\mathcal{M}_Q^*(P,r)$.*

PROOF. Most of these results are contained in [**KS04**], they correspond to 8.2.2, 8.2.3, 8.2.7, 8.2.11, 8.3.4, 8.4.2 and 8.4.3. Statement (6) follows from Thompson's $P \times Q$-Lemma, but here is a direct argument: Suppose that P centralises a centraliser closed subgroup Q_0 of Q and set $Q^* := C_Q(P)$. Then $Q_0 \leq Q^*$ and therefore Q^* is centraliser closed in Q. Moreover Q^* is subnormal in Q because Q is nilpotent. We argue by induction on $|Q|$ and therefore suppose that the result holds for all proper P-invariant subgroups of Q that contain Q^*. Now we note that $N_Q(Q^*)$ is P-invariant, so either $N_Q(Q^*) = Q$ and hence $Q^* \trianglelefteq Q$ or $N_Q(Q^*) < Q$ in which case we observe that P centralises the centraliser closed subgroup Q^* of $N_Q(Q^*)$. Then P centralises $N_Q(Q^*)$ by induction, so $N_Q(Q^*) \leq C_Q(P) \leq Q^*$ and $Q^* = Q$. Thus we look at the case where $Q^* \trianglelefteq Q$. As Q normalises Q^* and P centralises it, we see that $[Q,P]$ centralises Q^* and therefore
$$[Q,P] \leq C_Q(Q^*) \leq Q^* = C_Q(P).$$
Now $[Q,P,P] = 1$ and (2) yields that P centralises Q. □

LEMMA 2.2 (Thompson's $P \times Q$-Lemma). *Suppose that X acts on a p-group W and that $X = PQ$ is a central product of a p-group P and a p-perfect group Q. If Q centralises $C_W(P)$, then Q centralises W.*

PROOF. This is (24.2) in [**Asc00**]. □

LEMMA 2.3. *Suppose that p is an odd prime and that P is a p-group of rank at most 2. If q is a prime divisor of $|\mathrm{Aut}(P)|$ distinct from p, then $q < p$.*

PROOF. This is a combination of Lemmas 4.7 and 4.13 in [**BG94**]. □

LEMMA 2.4. *Suppose that X is a p-group. Then there exists a characteristic subgroup P of X (a so called **critical** subgroup) such that*

- *every p'-subgroup of $Aut(X)$ is faithful on P,*
- *$P' = \phi(P)$ is elementary abelian and lies in $Z(P)$ and*
- *if X is not abelian, then $exp(P) = p$ if p is odd and $exp(P) = 4$ if $p = 2$.*

PROOF. Proposition 11.11 in [**GLS96**]. □

LEMMA 2.5. *Let $H \leq X$ be a $2'$-subgroup that is normalised by an involution $t \in X$. Suppose that every t-invariant π-subgroup of H is centralised by t. Then $H = C_H(t)O_{\pi'}(H)$.*

PROOF. This is Lemma 2.2 in [**Wal09**]. □

LEMMA 2.6. *Suppose that q is odd and that an involution $t \in X$ acts on a q-subgroup Q of X. If $r(Q) \geq 3$, then Q possesses a t-invariant elementary abelian subgroup of order q^3.*

PROOF. This is Lemma 11.18 in [**GLS96**]. □

THEOREM 2.7. *Suppose that A and A_0 are groups, that A is an elementary abelian p-group of rank at least 3 and that the central product AA_0 acts coprimely on X. Suppose that X is soluble and that $X = [X, A_0]$. Furthermore, let $B \leq A$, let $H := C_X(A_0 B)$ and let $Hyp^2(A)$ denote the set of all subgroups of A of index p^2. Then*
$$H = \langle [C_X(Y), A_0] \cap H \mid Y \in Hyp^2(A) \rangle.$$

PROOF. This result is proved in [**Wal08**]. □

LEMMA 2.8. *Let Y be a p-subgroup of $O_{p',p}(X)$. Then $O_{p'}(C_X(Y)) \leq O_{p'}(X)$. If X is soluble and if Y is a p-subgroup of X, then $O_{p'}(C_X(Y)) \leq O_{p'}(X)$.*

PROOF. The second statement is often referred to as Goldschmidt's Lemma and can be found for example as Proposition 1.15(b) in [**BG94**]. We give an argument for the first statement. We let $Y \leq O_{p',p}(X)$ and we may suppose that $O_{p'}(X) = 1$, so that $Y \leq O_p(X)$ and hence $E(X) \leq C_X(Y)$. Let $Q := O_{p'}(C_X(Y))$. Then
$$[E(X), Q] \leq E(X) \cap Q \leq O_{p'}(E(X)) \leq O_{p'}(X) = 1.$$
Moreover $Y \times Q$ acts on $O_p(X)$ and $[C_{O_p(X)}(Y), Q] \leq O_{p'}(X) \cap Q = 1$. Then Thompson's $P \times Q$-Lemma 2.2 yields that Q centralises $O_p(X)$. We recall that $O_{p'}(X) = 1$ and therefore $F^*(X) = O_p(X)E(X)$. It follows that
$$Q \leq C_X(F^*(X)) = Z(F(X)) \leq O_p(X)$$
and hence $Q = 1$. □

LEMMA 2.9. *Let $t \in O_{2',2}(X)$ be an involution and let $D \leq X$ be a $C_X(t)$-invariant $2'$-subgroup. Then $D \leq O(X)$. If D is nilpotent, then $[D, t] \leq F(X)$.*

2.2. GENERAL RESULTS

PROOF. Lemma 3.6 in [**Wal09**]. □

LEMMA 2.10. *Let V be a p-subgroup of $O_{p',p}(X)$. Then for all $q \in p'$, the subgroup $C_X(V)$ is transitive on the set $\mathcal{W}_X^*(V,q)$.*

PROOF. Lemma 3.7 in [**Wal09**]. □

LEMMA 2.11. *Suppose that X is a 2-group and that $X_0 \trianglelefteq X$. If $r(X_0) \geq 2$, then either X_0 contains a normal elementary abelian subgroup of X of order 4 or X_0 is dihedral or semi-dihedral.*

PROOF. Lemma 10.11 in [**GLS96**]. □

LEMMA 2.12. *Suppose that X is a 2-group and that $t \in X$ is an involution. If $C_X(t)$ is elementary abelian of order 4, then X is dihedral or semi-dihedral.*

PROOF. This is 5.3.10 in [**KS04**]. □

LEMMA 2.13. *Suppose that $Q_1, Q_2 \leq X$ are commuting quaternion groups and that $Z(Q_1) = Z(Q_2)$. Then the rank of $Q_1 Q_2$ is 3.*

PROOF. Let $Q := Q_1 Q_2$ and $Z := Z(Q_1)$. Let $a_1, a_2, b_1, b_2 \in Q$ be such that a_1 and a_2 have 2-power order, that b_1, b_2 have order 4 and that $Q_1 = \langle a_1, b_1 \rangle$ and $Q_2 = \langle a_2, b_2 \rangle$. Then some powers c_1 of a_1 and c_2 of a_2 have order 4. Our hypothesis $Z = Z(Q_2)$ implies that the unique involution s in Q_1 is also the unique involution in Q_2. In particular the elements b_1, b_2, c_1, c_2 all have the same square, namely s. As Q_1 and Q_2 centralise each other, it follows that $(b_1 b_2)^2 = s^2 = 1$. Therefore $b_1 b_2$ and similarly $c_1 c_2$ are diagonal involutions. They are distinct from each other and distinct from s, in fact $A := \langle b_1 b_2, c_1 c_2, s \rangle$ is an elementary abelian subgroup of $Q_1 Q_2$ of order 8 and hence $r(Q) \geq 3$.

Now let a be an arbitrary involution in $Q \backslash Z$. Then there are elements $x_1 \in Q_1$ and $x_2 \in Q_2$ of order 4, respectively, such that $a = x_1 x_2$. The subgroups $C_{Q_1}(x_1)$ and $C_{Q_2}(x_2)$ are cyclic of order at least 4 with intersection Z, so $C_{Q_1}(x_1) C_{Q_2}(x_2)$ has rank 2. If, for $i \in \{1,2\}$, we let $y_i \in Q_i \backslash C_{Q_i}(x_i)$ be an element of order 4, then $y_1 y_2$ is an involution that commutes with $C_{Q_1}(x_1) C_{Q_2}(x_2)$, but is not contained in it. As $C_Q(a) = C_{Q_1}(x_1) C_{Q_2}(x_2) \langle y_1 y_2 \rangle$, this implies that $C_Q(a)$ has rank 3 and thus $r(Q) = 3$. □

LEMMA 2.14. *Suppose that X is a dihedral group of order at least 8 or a semi-dihedral group of order at least 16. Then $Aut(X)$ is a 2-group.*

PROOF. This is Proposition 4.53 in [**Cra11**]. □

THEOREM 2.15. *Suppose that X is a 2-group with precisely three involutions. Then $Aut(X)$ is soluble.*

PROOF. This is Theorem 3.16 in [**Wal09**]. □

THEOREM 2.16. *Suppose that X is quasi-simple and that $n \in \mathbb{N}$ is such that $n \geq 5$ and $X/Z(X) \simeq A_n$.*

(1) *If $n \neq 6$, then $Aut(X) \simeq \mathcal{S}_n$.*
(2) *If $Z(X) = 1$, then $r_2(X) = 2 \cdot k$, where k is the largest integer less than or equal to $\frac{n}{4}$.*
(3) *If 2 divides $|Z(X)|$, then $r_2(X) = 3 \cdot l + 1$, where l is the largest integer less than or equal to $\frac{n}{8}$.*

PROOF. Theorem 5.2.1 and Proposition 5.2.10 in [**GLS98**]. □

LEMMA 2.17. *Let $S \in Syl_2(X)$ and suppose that P and Q are subgroups of S such that $S = PQ$. Suppose that $P \trianglelefteq S$, that Q is cyclic and that $P \cap Q = 1$. Let s be the unique involution in Q. If $O^2(X) = X$, then there exists an X-conjugate t of s in P such that $C_S(t) \in Syl_2(C_X(t))$.*

PROOF. This is a generalisation of Thompson's Transfer Lemma and can be found as Lemma 15.16 in [**GLS96**]. □

LEMMA 2.18. *Suppose that $H \leq X$ contains a Sylow p-subgroup of X. If for every p-element $y \in H$ we have that $y^X \cap H = y^H$, then $O^p(X) = X$ if and only if $O^p(H) = H$.*

PROOF. Lemma 15.10 (ii) in [**GLS96**]. □

THEOREM 2.19. *Suppose that $S \in Syl_2(X)$ is cyclic. Then $X = SO(X)$. In particular the unique involution in S is contained in $Z^*(X)$.*

PROOF. This is 7.2.2 in [**KS04**] and follows from Burnside's p-Complement Theorem. □

THEOREM 2.20 (Brauer-Suzuki). *Suppose that $S \in Syl_2(X)$ is quaternion and let s be the unique involution in S. Then $s \in Z^*(X)$.*

PROOF. See [**Gla74**] for a proof using ordinary character theory. □

THEOREM 2.21 (Gorenstein-Walter). *Suppose that X is non-abelian and simple and has dihedral Sylow 2-subgroups. Then X is isomorphic to A_7 or there exists an odd number $q \geq 5$ such that $X \simeq PSL_2(q)$.*

PROOF. See [**Ben81**] and [**BG81**] for a proof using ordinary character theory. □

LEMMA 2.22. *Suppose that E is a component of X and that $S \in Syl_2(E)$ is of rank 1. Then E is not simple and $E/Z(E)$ is isomorphic to A_7 or there exists an odd number $q \geq 5$ such that $E/Z(E) \simeq PSL_2(q)$.*

PROOF. We first recall that, since E is a component (and hence quasi-simple), we have that $O(E) \leq Z(E)$. Now we look at S. This is a 2-group of rank 1 and hence S is cyclic or quaternion. By Theorem 2.19 it follows that, in the cyclic case, $E = O(E)S$. But $O(E) \leq Z(E)$ and consequently $E/Z(E)$ is a 2-group, which is a contradiction. Therefore S is quaternion and the Brauer-Suzuki Theorem 2.20 implies that the unique involution in S lies in $Z(E)$. In particular E is not simple. Moreover a Sylow 2-subgroup of $E/Z(E)$ is dihedral and the Gorenstein-Walter Theorem 2.21 yields the result. □

THEOREM 2.23. *Suppose that X has odd order and let $P \in Syl_p(X)$. If $char(X) = p$, then $ZJ(P) \trianglelefteq X$.*

PROOF. This is a weakened version of Glauberman's ZJ-Theorem (in [**Gla68a**]). □

THEOREM 2.24. *Suppose that X is a π-group and that A is a π'-group of automorphisms of X. Suppose that t is an automorphism of X of order 2 such that $C_X(t) \leq C_X(A)$. Then $[C_X(A), t]$ and $[X, A]$ are normal subgroups of X and $[X, A]$ is nilpotent of odd order.*

PROOF. This is a weakened version of Theorem 1 in [**Gla72**]. □

We state special cases of the main results in [**Wal11**] for reference. They will be applied later in order to exclude cases where maximal subgroups of characteristic p contain centralisers of involutions. One definition is necessary.

DEFINITION 2.25. Suppose that H_1, H_2 are distinct proper subgroups of X and that $V \leq H_1 \cap H_2$. Let $\pi := \pi(V)$ and $q \in \pi'$. Then we say that the pair (H_1, H_2) is a **V-special primitive pair of characteristic q of X** if and only if the following hold:

- H_1 and H_2 are primitive;
- for all $i \in \{1, 2\}$, we have that $F^*(H_i) = O_q(H_i)$;
- $C_X(V) \leq H_1 \cap H_2$ and
- $V \leq Z^*_\pi(H_1) \cap Z^*_\pi(H_2)$.

THEOREM 2.26. *Suppose that V is an abelian 2-subgroup of X. Suppose that q is odd, that $O_q(X) = 1$ and that, whenever $C_X(V) \leq H < X$, then $\widehat{H} := H/O(H)$ has a unique maximal $\widehat{C_X(V)}$-invariant q-subgroup. If (H_1, H_2) is a V-special primitive pair of characteristic q of X, then $O_q(H_1) \cap H_2 = 1 = O_q(H_2) \cap H_1$.*

PROOF. This follows from Theorem I in [**Wal11**]. □

THEOREM 2.27. *Suppose that $a \in X$ is an involution. Suppose that q is odd, that $O_q(X) = 1$ and that, whenever $C_X(a) \leq H < X$, then $a \in Z^*(H)$. If (H_1, H_2) is an $\langle a \rangle$-special primitive pair of characteristic q of X, then $O_q(H_1) \cap H_2 = 1 = O_q(H_2) \cap H_1$.*

PROOF. This follows from Theorem II in [**Wal11**]. □

2.3. A Nilpotent Action Result

THEOREM 2.28. *Let p be an odd prime, let $\mathbb{F}$ be a field of characteristic p and V a finite dimensional $\mathbb{F}$-vector space. Suppose that X acts on V and that $t \in Z^*(X)$ is an involution such that $\dim(C_V(t)) \leq 1$. If $X/O(X)$ is 2- and 3-perfect, then the action of $[X,t]$ on V is nilpotent.*

PROOF. We first notice that $[X,t] \leq O(X)$ because $t \in Z^*(X)$. It follows that $X = C_X(t)O(X)$ and that $[X,t]$ is soluble because it has odd order. Assume that the theorem is false and choose X to be a minimal counter-example. More precisely we choose X such that $X/O(X)$ is 2- and 3-perfect and that the action of $[X,t]$ on V is not nilpotent, and we assume that $|X| + \dim(V)$ is as small as possible. We may suppose that $\mathbb{F}$ is algebraically closed. It follows that X acts faithfully on V. Next we show that the action of X is irreducible and that, consequently, $O_p(X) = 1$:

If W is a proper X-invariant subspace of V, then by induction $[X,t]$ acts nilpotently on W and on V/W and hence on V, which is a contradiction. Thus X acts irreducibly.

As the action of $[X,t]$ on V is not nilpotent by assumption, we have that $[X,t] \neq 1$ and in particular $[O(X),t] \neq 1$. Therefore t does not centralise $F(O(X))$ and we find an odd prime q such that $[O_q(X),t] \neq 1$. Applying Lemma 2.4, let Q be a critical subgroup of $O_q(X)$. Then $[Q,t] \neq 1$ and $\Phi(Q) \leq Z(Q) =: Z$ (by Lemma 2.4). Moreover Q is abelian or of exponent q. We note that $Q \trianglelefteq X$ and therefore $C_V(Q) = 0$.

(1) Let $n \in \mathbb{N}$ and $t_1, ..., t_n \in t^Q$. Let $Q_0 := Q \cap \langle t_i \mid i \in \{1, ..., n\}\rangle$. Then $\dim([V, Q_0]) \leq n$.

   PROOF. Let $i \in \{1, ..., n\}$. Then $\dim(C_V(t_i)) = \dim(C_V(t)) \leq 1$ and therefore $[V, t_i]$ has codimension at most 1.

   Let $U := \bigcap_{i \in \{1,...,n\}} [V, t_i]$. Then U has codimension at most n in V and every element in $\{t_1, ..., t_n\}$ inverts U. As Q has odd order, we see that Q_0 is generated by products of an even number of conjugates of t and therefore Q_0 centralises U. As U has codimension at most n, this implies that $\dim([V, Q_0]) \leq n$. □

(2) Let $\overline{X} := X/O(X)$. Then $\overline{C_X(t)} = O^2(\overline{C_X(t)}) = O^3(\overline{C_X(t)})$.

   PROOF. We know that $\overline{X} = O^2(\overline{X}) = O^3(\overline{X})$ by hypothesis. As $t \in Z^*(X)$ and hence $\overline{X} = \overline{C_X(t)}$, this implies the statement. □

(3) $[Z,t] = 1$. In particular Q is not abelian and hence Q has exponent q. (So Q is extra-special.)

   PROOF. Assume otherwise and let $y \in Z^{\#}$ be such that y is inverted by t. Then $\langle t, y \rangle = \langle t, t^y \rangle$ and (1) yields that $\dim([V, y]) \leq 2$. Let V_1 and V_2 denote the distinct 1-dimensional eigenspaces for y on V and let $V_3 := [V,t] \cap [V, t^y]$. Then $V_3 = C_V(y)$ and $V = V_1 \oplus V_2 \oplus V_3$. Moreover Q normalises V_1, V_2 and V_3 and t inverts V_3, therefore $[Q,t]$ centralises V_3. It follows that $[V, [Q,t]] = [V_1 \oplus V_2, [Q,t]] = V_1 \oplus V_2 = [V, y]$ and hence $[Q,t] = \langle y \rangle$. In particular $\langle y \rangle$ is $C_X(t)$-invariant, but not centralised by

$C_X(t)$ because t inverts y. This implies that $C_{C_X(t)}(y)$ has even index in $C_X(t)$, contrary to (2). As $[Q,t] \neq 1$, it follows that Q is not abelian. □

(4) $q = 3$. If $y_1, y_2 \in Q^\#$ are inverted by t and if $R := \langle y_1, y_2 \rangle$ is extra-special of order q^3, then $\dim([V,R]) = 3$ and $R = [Q,t]$.

PROOF. Suppose that $y_1, y_2 \in Q^\#$ are inverted by t and that $R := \langle y_1, y_2 \rangle$ is extra-special of order q^3. Then $R\langle t \rangle = \langle t, t^{y_1}, t^{y_2} \rangle$ and therefore $\dim([V,R]) \leq 3$ by (1). As $[V,R]$ is a faithful $\mathbb{F}R$-module and q is odd, it follows that $\dim([V,R]) \geq q \geq 3$ and so $q = 3 = \dim([V,R])$. Then R acts irreducibly on $[V,R]$ and we have proved most of the results in (4). For the last assertion let $x \in Z(R)$ be such that $[V,R] = [V,x]$. We assume that $R \neq [Q,t]$. Then there exists an element $y_3 \in Q \setminus R$ of order q that is inverted by t. We set $Y := R\langle y_3 \rangle$ and apply (1), so that $\dim([V,Y]) \leq 4$. The subgroup $P := \langle y_1, y_3 \rangle$ of Q is extra-special of order 27 with $Z(P) = Z = Z(R)$, because Q is extra-special by (3) and $y_1, y_3 \notin Z$. Let $x' \in Z(P)$ be such that $[V,P] = [V,x']$ and let $U := [V,x] \cap [V,x']$. Then $U = [V,P] \cap [V,R]$ is 2-dimensional because $\dim([V,P] + [V,R]) = \dim([V,Y]) \leq 4$ and $[V,P] \neq [V,R]$. But x and x' lie in $Z(R)$ and therefore R normalises $[V,x] \cap [V,x']$. This is impossible because this is a proper subspace of $[V,R]$ and R acts irreducibly on $[V,R]$. Thus $R = [Q,t]$ and the proof of (4) is complete. □

Let $y_1, y_2 \in Q^\#$ be distinct such that t inverts y_1 and y_2 and such that $\langle y_1, y_2 \rangle$ is extra-special of order 27. Let $R := [Q,t]$ and $x \in Z(R)^\#$. Then $R = \langle y_1, y_2 \rangle$ by (4) and $C_X(t)$ normalises $\langle x \rangle$ and therefore acts on $R/\langle x \rangle$. This group is elementary abelian of order 9 and $C_X(t)$ normalises it, so $C_X(t)/C_{C_X(t)}(R/\langle x \rangle)$ is isomorphic to a subgroup of $GL_2(3)$. This implies that $C_X(t)/C_{C_X(t)}(R/\langle x \rangle)$ has a non-trivial 2- or 3-factor group whence the same holds for $C_X(t)$. This contradicts (2) and hence the proof of the theorem is complete. □

REMARK 2.29. The non-split extension X of 3^{1+2} with $SL_2(3)$ acting on it (i.e. X is a non-3-perfect $\{2,3\}$-group) arises naturally in the proof. Considering the action of X on a 3-dimensional vectorspace over $\mathrm{GF}(7)$ illustrates why a more general result, namely omitting the hypothesis that X is 3-perfect, does not hold.

COROLLARY 2.30. *Suppose that $p \in \pi(X)$ is odd and let $P := O_p(X)$. Suppose that $X/O(X)$ is 2- and 3-perfect and that $t \in Z^*(X)$ is an involution such that $C_P(t)$ is cyclic. Then the action of $[X,t]$ on P is nilpotent.*

PROOF. This follows from Theorem 2.28 because X acts on the elementary abelian p-group $P/\Phi(P)$. □

CHAPTER 3

Isolated Involutions

From now on G is a finite group and $z \in G$ is an isolated involution. We set $C := C_G(z)$ and start by collecting some basic facts. Then we deduce knowledge about the set $K := \{zz^g \mid g \in G\}$ of commutators and use it to make initial statements about the structure of G.

LEMMA 3.1. *Let* $z \in S \in Syl_2(G)$.
 (1) $z^G \cap S = \{z\}$.
 (2) *Every 2-subgroup of G that is normalised by z is centralised by z. In particular $z \in Z(S)$.*
 (3) *For all $g \in G$, the element zz^g has odd order.*
 (4) *Whenever $z \in H \le G$, then $z^G \cap H = z^H$.*
 (5) *Let $w \in G\backslash z^G$ be an involution. Then the order of zw is even, but not divisible by 4. In particular, the Sylow 2-subgroups of $\langle z, w\rangle$ are elementary abelian of order 4.*
 (6) *If $z \in X \trianglelefteq Y \le G$, then $Y = XC_Y(z)$.*
 (7) *Suppose that $z \notin N \trianglelefteq G$ and let $\overline{G} := G/N$. Then $C_{\overline{G}}(\overline{z}) = \overline{C}$ and $\overline{z}$ is isolated in $\overline{G}$.*
 (8) *If $C \le H \le G$, then H is the only conjugate of H in G that contains z.*
 (9) *If $H \le G$ is a z-invariant subgroup, then $H \cap C$ controls fusion in $H \cap C$ with respect to H.*
 (10) $O^2(G) = G$ *if and only if* $O^2(C) = C$.
 (11) *If $s, t \in z^G$ are distinct, then $st \notin C$.*

PROOF. (1)-(3) are straightforward from the definition of "isolated".
 (4) Let $g \in G$ be such that $z^g \in H$. We observe that $\langle z, z^g\rangle$ is a dihedral group of twice odd order by (3). Thus z and z^g are conjugate in $\langle z, z^g\rangle$ by Sylow's Theorem.
 (5) Set $D := \langle z, w\rangle$ and note that zw has even order because otherwise z and w are conjugate. Let $z \in T \in Syl_2(D)$. Then $z \in Z(T)$ by (2) and on the other hand a power of zw is the unique central involution in D. Therefore T is elementary abelian of order 4.
 (6) Let $z \in P \in Syl_2(X)$. As z is isolated and central in P by (2), we have that $N_Y(P) \le C_Y(z)$. Hence with a Frattini argument, it follows that $Y = XN_Y(P) \le XC_Y(z)$ as stated.
 (7) Of course $\overline{C} \le C_{\overline{G}}(\overline{z})$, so now we prove the converse. It follows from (4) that N acts transitively on $z^G \cap N$, so every z-invariant coset of N in G has a representative from C. Therefore $C_{\overline{G}}(\overline{z}) \le \overline{C}$ and the second statement follows from there.

(8) Assume that $g \in G \backslash N_G(H)$ is such that $z \in H^g$. Then $z \in H \cap H^g$ and therefore $z, z^{g^{-1}} \in H$. It follows from (4) that there exists an element $h \in H$ such that $z = z^{hg}$. Hence $hg \in C \leq H$ and thus $g \in H$, which is a contradiction.

(9) Let $x, y \in H \cap C$ and let $h \in H$ be such that $x^h = y$. As x, x^h are both contained in C, it follows that $z, z^{h^{-1}} \in C_H(x)\langle z \rangle$. But then (4) yields that z and $z^{h^{-1}}$ are conjugate in $C_H(x)\langle z \rangle$. Let $a \in C_H(x)\langle z \rangle$ be such that $z^a = z^{h^{-1}}$ and note that $z^{ah} = z$ with this choice. As $C_H(x)$ is z-invariant, we find some $b \in C_H(x)$ such that $a = zb$ and we see that $z^{bh} = z^{zbh} = z^{ah} = z$. This means that $bh \in C \cap H$ and $x^{bh} = x^h = y$.

(10) By (2) we know that C contains a Sylow 2-subgroup of G. Now suppose that $y \in C$ is a 2-element. Then $y^G \cap C = y^C$ by (9). So the result follows from Lemma 2.18.

(11) Assume that $st \in C$ and set $X := C_G(st)$ and $Y := X\langle t \rangle$. Then t inverts st and z centralises st and thus $z \in X \triangleleft Y$ and $t \notin X$. But z and t are both contained in Y and therefore conjugate in Y by (4). This is impossible. □

DEFINITION 3.2. Recall that $K = \{zz^g \mid g \in G\}$. We define an operation $\circ$ in the following way: For all $a, b \in K$ we set $a \circ b := aba$.

Fischer introduced such an operation in a more general context in [**Fis64**] where he proves a special case of the Z*-Theorem. Glauberman refers to Fischer's result in [**Gla66a**] and he mentions in [**Gla68b**] that the Z*-Theorem is a group theoretic equivalent to the fact that certain finite loops of odd order – which he refers to as B-loops – are soluble. Therefore the following construction will look familiar to any reader who has seen the corresponding results from loop theory.

LEMMA 3.3.
  (1) K is C-invariant and contains 1.
  (2) An element $x \in G$ is contained in K if and only if x has odd order and z inverts x.
  (3) Let $a \in K$. Then for all $n \in \mathbb{N}$, the element a^n lies in K.
  (4) $\circ$ is a binary operation on K.
  (5) Let $a, b, d \in K$. If $a \circ b = d$, then $a^{-1} \circ d = b$. Moreover $a^{-1} \circ b^{-1} = (a \circ b)^{-1}$.
  (6) For all $a \in K$, the maps $k \mapsto k \circ a$ and $k \mapsto a \circ k$ are bijective on K.

PROOF. The first statement is immediate. For the remainder let $a, b \in K$ be arbitrary and let $g, h \in G$ be such that $a = zz^g$ and $b = zz^h$. Then $\langle z, z^g \rangle$ is a dihedral group of order $2 \cdot o(a)$ and $o(a)$ is odd by Lemma 3.1 (3), moreover z inverts a. Conversely suppose that $x \in G$ has odd order and is inverted by z. Then $x = z \cdot zx$ and $zx \in z^{\langle x \rangle}$, therefore $x \in K$ and (2) holds. For (3) we observe that z inverts a and hence it inverts a^n, so as a^n has odd order it follows that $a^n \in K$ by (2). Looking at (4) we calculate

$$a \circ b = aba = zz^g zz^h zz^g = zz^{ha} \in K$$

and therefore $\circ$ is a binary operation on K.

For (5) we recall that $a \circ b = d$ means that $aba = d$. Thus $a^{-1} \circ d = a^{-1}da^{-1} = b$ as stated. Finally

$$(a \circ b)^{-1} = (aba)^{-1} = a^{-1}b^{-1}a^{-1} = a^{-1} \circ b^{-1}.$$

In (6) it suffices to show that both maps are injective on K. Let $d \in K$ and let $k \in G$ be such that $d = zz^k$. Suppose that $a \circ b = a \circ d$. Then immediately $b = d$. Now if $a \circ b = d \circ b$, then $zz^g zz^h zz^g = zz^k zz^h zz^k$ and it follows that $z^{ha} = z^{hd}$. Hence $had^{-1}h^{-1} \in C$ and this means that $z^{gzh^{-1}} z^{kzh^{-1}} \in C$. Then Lemma 3.1 (11) forces $z^{gzh^{-1}} = z^{kzh^{-1}}$ and therefore $gzh^{-1}hzk^{-1} \in C$. This yields that $gk^{-1} \in C$, so $z^g = z^k$ and finally $a = d$. □

DEFINITION 3.4. For all $a, b \in K$, we denote by $a + b$ the (by Lemma 3.3 (6)) unique element d in K with the property that $d \circ a^{-1} = b$. In other words, $(a+b)a^{-1}(a+b) = (a+b) \circ a^{-1} = b$.

LEMMA 3.5. Let $a, b, d \in K$.
(1) $a + b = b + a$.
(2) For all $c \in C$ we have that $(a+b)^c = a^c + b^c$.
(3) $(a+b)^{-1} = a^{-1} + b^{-1}$.
(4) $a + b = 1$ if and only if $b = a^{-1}$.
(5) $a \circ (b+d) = a \circ b + a \circ d$.

PROOF. We have that $(a+b) \circ a^{-1} = b$ by definition. Lemma 3.3 (5) yields that $(a+b)^{-1} \circ b = a^{-1}$ and then $(a+b) \circ b^{-1} = a$. But Definition 3.4 implies that $a = (b+a) \circ b^{-1}$ and hence that $a + b = b + a$.

For all $c \in C$ we know that c acts on K by Lemma 3.3 (1). Also, by Definition 3.4, we see that $(a+b)a^{-1}(a+b) = b$ and therefore $b^c = (a+b)^c (a^{-1})^c (a+b)^c$. Consequently (2) holds. Then (3) follows from (2) because z is in C and inverts K, by Lemma 3.3 (2). For (4) we see, just using Definition 3.4, that $a + b = 1$ if and only if $1 \circ a^{-1} = b$, and this holds if and only if $a^{-1} = 1a^{-1}1 = b$. For the last assertion we recall that $(b+d) \circ b^{-1} = d$ by definition. This gives that $a \circ d = a \circ ((b+d) \circ b^{-1})$. On the other hand, by definition of the element $a \circ b + a \circ d$, we have that $a \circ d = (a \circ b + a \circ d) \circ (a \circ b)^{-1}$. This yields that

$(a \circ b + a \circ d) \circ (a \circ b)^{-1} = a \circ d = a \circ ((b+d) \circ b^{-1}) = a((b+d)b^{-1}(b+d))a$
$= a((b+d)aa^{-1}b^{-1}a^{-1}a(b+d))a = a(b+d)a((a \circ b)^{-1})a(b+d)a,$

by Lemma 3.3 (5). But

$a(b+d)a((a \circ b)^{-1})a(b+d)a = (a \circ (b+d))(a \circ b)^{-1}(a \circ (b+d)) = (a \circ (b+d)) \circ (a \circ b)^{-1},$

therefore

$$(a \circ b + a \circ d) \circ (a \circ b)^{-1} = (a \circ (b+d)) \circ (a \circ b)^{-1}$$

and Lemma 3.3 (6) gives the result. □

THEOREM 3.6. Let $a \in K$ and let $s \in C$ be an involution. Then there exist elements $u \in C_K(s)$ and $v \in C_K(sz)$ such that $a = u \circ v$, and this representation of a is unique. In particular $|K| = |C_K(s)||C_K(sz)|$ and $K \subseteq \langle C_K(s), C_K(sz) \rangle$.

PROOF. Lemma 3.5 (1) and (2) imply that $a + a^s = a^s + a = (a + a^s)^s$ and therefore $a + a^s \in C_K(s)$.

Now, for all $b \in K$, we define $\bar{b} := b + b^s$ and we set $J := \{b \in K \mid \bar{b} = 1\}$. Then Lemma 3.5 (4) yields that

$$J = \{b \in K \mid b + b^s = 1\} = I_K(s) = C_K(sz).$$

As $\bar{a} \in K$ is of odd order (Lemma 3.3 (2)), there exists a power y of $\bar{a}$ with the property that $(y^{-1})^2 = \bar{a}$. We pick this element y and observe that, by Lemma 3.3 (3), it is contained in K and thus lies in $C_K(s)$. Furthermore $y \circ \bar{a} = 1$. Lemma 3.5 (5) and the fact that s centralises y imply that
$$y \circ \bar{a} = y \circ (a + a^s) = y \circ a + y \circ a^s = y \circ a + (y \circ a)^s = \overline{y \circ a}.$$
Thus $\overline{y \circ a} = y \circ \bar{a} = 1$ which means that $y \circ a \in J$. Now let $u := y^{-1}$ and $v := y \circ a$. Then
$$a = y^{-1}yayy^{-1} = y^{-1} \circ (y \circ a) = u \circ v \in C_K(s) \circ C_K(sz).$$
This proves the existence of a representation as stated.

For the uniqueness we suppose that $u' \in C_K(s)$ and $v' \in C_K(sz)$ are such that $a = u' \circ v'$. Then
$$\bar{a} = \overline{u' \circ v'} = (u' \circ v') + (u' \circ v')^s = (u' \circ v') + (u' \circ v'^s) = u' \circ (v' + v'^s)$$
where the last equality comes from Lemma 3.5 (5). Moreover $v' \in J = I_K(s)$ by choice which implies that $\overline{v'} = 1$. We deduce that
$$\bar{a} = u' \circ (v' + v'^s) = u' \circ \overline{v'} = u' \circ 1 = (u')^2$$
and therefore $(u')^2 = \bar{a} = u^2$. As u and u' are of odd order, we obtain that $u = u'$. Finally Lemma 3.3 (6) yields that also $v = v'$. $\square$

LEMMA 3.7. *Suppose that $z \in H \leq G$. Then $H = C_H(z)(H \cap K)$. More precisely, every coset of $C_H(z)$ in H contains a unique element that is inverted by z. In particular we have that $G = CK$ and that every coset of C in G contains a unique element that is inverted by z. Moreover, for every involution $s \in C$, we have that $|G| = |C||C_K(sz)||C_K(s)|$.*

PROOF. Set $C_0 := C_H(z)$. As K is C-invariant, it follows that $H \cap K$ is C_0-invariant and every non-trivial element in $H \cap K$ is inverted and not centralised by z. Therefore $(H \cap K) \cap C_0 = 1$. We also know that $|H : C_0| \leq |H \cap K|$ because $\{zz^h \mid h \in H\} \subseteq H \cap K$. Now we show that $H \cap K$ contains a unique representative for every coset of C_0 in H. Suppose that $zz^g, zz^h \in H \cap K$ are such that $C_0 zz^g = C_0 zz^h$. Then $z^g z^h \in C_0 \leq C$ which by Lemma 3.1 (11) is only possible if $z^g = z^h$.

The first two statements for G follow from this and, together with Theorem 3.6, this implies that $|G| = |C||C_K(sz)||C_K(s)|$ as stated. $\square$

LEMMA 3.8. *Let $p \in \pi(G)$. Then $\mathcal{W}_G^*(\langle z \rangle, p) \subseteq Syl_p(G)$.*

PROOF. As z lies in a Sylow 2-subgroup of G, we only need to discuss the case that p is odd. We proceed by induction on $|G|$ and first show that $\mathcal{W}_G(\langle z \rangle, p) \neq \{1\}$. Suppose that $r_2(G) = 1$. Then the Sylow 2-subgroups of G are cyclic or quaternion. It follows that $z \in Z^*(G)$ by Theorem 2.19 or the Brauer-Suzuki Theorem 2.20, respectively. But then $G = CO(G)$ and at least one of these subgroups has order divisible by p. If p divides $|C|$, then z centralises a non-trivial p-subgroup of G. If p divides $|O(G)|$, then Lemma 2.1 (7) yields that $\{1\} \neq \mathcal{W}_{O(G)}(\langle z \rangle, p) \subseteq \mathcal{W}_G(\langle z \rangle, p)$.

Thus we may suppose that $r_2(G) \geq 2$ and we choose an involution $s \in C$ distinct from z. By Lemma 3.7, the prime p divides one of $|C|$, $|C_K(s)|$ or $|C_K(sz)|$. If p divides $|C|$, then there is nothing left to prove. Suppose therefore that p does not divide $|C|$. Then by Lemma 3.7 and by symmetry between s and sz we may suppose that p divides $|C_K(s)|$. If $C_G(s) < G$, then $\mathcal{M}_{C_G(s)}(\langle z \rangle, p) \neq \{1\}$ by induction because z is contained in $C_G(s)$. If $C_G(s) = G$, then $s \in Z(G)$. We can therefore argue by induction in the factor group $G/\langle s \rangle$, applying Lemma 3.1 (7). We conclude that $\mathcal{M}_G(\langle z \rangle, p) \neq \{1\}$.

Now let $P_0 \in \mathcal{M}_G^*(\langle z \rangle, p)$ and let $N_0 := N_G(P_0)$. Then we have that $z \in N_0$. First suppose that $N_0 < G$. Then induction yields that $\mathcal{M}_{N_0}^*(\langle z \rangle, p) \subseteq \mathrm{Syl}_p(N_0)$. By the maximal choice of P_0, this implies that $P_0 \in \mathrm{Syl}_p(N_0)$ and therefore that $P_0 \in \mathrm{Syl}_p(G)$. Now suppose that $N_0 = G$. Then $P_0 \trianglelefteq G$ and in G/P_0 there exists a z-invariant Sylow p-subgroup by induction, because $P_0 \neq 1$. Its pre-image in G is a z-invariant Sylow p-subgroup of G and equals P_0 by the maximal choice of P_0. $\square$

DEFINITION 3.9. From now on, for every subgroup H of G and for every prime p, we denote by $\mathrm{Syl}_p(H, z)$ the set of all z-invariant Sylow p-subgroups of H. Similarly, if V is a 2-subgroup of G, then we denote by $\mathrm{Syl}_p(H, V)$ the set of V-invariant Sylow p-subgroups of H.

LEMMA 3.10. *Let $p \in \pi(G)$. Then C acts transitively on $\mathrm{Syl}_p(G, z)$.*

PROOF. Let $P_1, P_2 \in \mathrm{Syl}_p(G, z)$ and let $g \in G$ be such that $P_1^g = P_2$. Since $z \in N_G(P_2) = (N_G(P_1))^g$, we conclude that z and z^g are both contained in $N_G(P_2)$. They are therefore conjugate in $N_G(P_2)$ by Lemma 3.1 (4). Choose $h \in N_G(P_2)$ such that $z = z^{gh}$. Then $gh \in C$ and $P_1^{gh} = P_2^h = P_2$. $\square$

LEMMA 3.11. *Let $V \leq G$ be an elementary abelian subgroup of order 4 that contains z and that is generated by (necessarily non-conjugate) isolated involutions. Let $p \in \pi(G)$. Then $\mathcal{M}_G^*(V, p) \subseteq \mathrm{Syl}_p(G)$ and $C_G(V) = N_G(V)$ is transitive on $\mathrm{Syl}_p(G, V)$.*

PROOF. We denote the involutions in V by z, a and b and we note that all previous results on isolated involutions can be applied to all these involutions. For example, we may apply Lemma 3.8 and arguments from its proof. Let p be a prime. With Sylow's Theorem we may suppose that p is odd. The first step is to show that $\mathcal{M}_G(V, p) \neq \{1\}$:

Lemma 3.7 yields that p divides $|C|$, $|C_K(a)|$ or $|C_K(b)|$. If p divides $|C|$, then with Lemma 3.8, applied to C and the isolated involution a, we see that $\mathcal{M}_C^*(V, p) \neq \{1\}$ and hence $\mathcal{M}_G^*(V, p) \neq \{1\}$. Therefore we may suppose that p divides $|C_K(a)| = |C_G(a) : C_C(a)|$. Then p divides $|C_G(a)|$ and therefore Lemma 3.8, applied to $C_G(a)$ and the isolated involution z, yields that $\mathcal{M}_{C_G(a)}^*(V, p) \neq \{1\}$. We deduce that $\mathcal{M}_G(V, p) \neq \{1\}$.

For the remainder of the proof we argue by induction on $|G|$. Let $P \in \mathcal{M}_G^*(V, p)$ and let $H := N_G(P)$. Then $V \leq H$. If $H < G$, then since a, b and z are isolated in H we may apply induction and we see that $\mathcal{M}_H^*(\langle z \rangle, p) \subseteq \mathrm{Syl}_p(H)$. Then the maximal choice of P implies that $P \in \mathrm{Syl}_p(H)$ and therefore that $P \in \mathrm{Syl}_p(G)$. If $H = G$, then $P \trianglelefteq G$ and in G/P there exists a V-invariant Sylow p-subgroup, again by induction and because we know that $P \neq 1$ from the previous paragraph. A pre-image of a V-invariant Sylow p-subgroup of G/P in G is a V-invariant Sylow p-subgroup of G and equals P by the maximal choice of P. This finishes the proof. $\square$

LEMMA 3.12. *Suppose that $V \leq G$ is elementary abelian of order 4 and that $z \in V$. Let a, b, z denote the involutions in V. Let p be a prime and suppose that $P \in Syl_p(G)$ is such that $P \leq C_G(a)$. Suppose that C does not contain any Sylow p-subgroup of G. Then $|C_K(b)|_p = 1$ and $|K|_p = |C_K(a)|_p \neq 1$.*

PROOF. From Lemma 3.7 we know that
$$|G|_p = |C|_p |C_K(a)|_p |C_K(b)|_p.$$
From our hypothesis we deduce that $|G|_p = |P| = |C_G(a)|_p$. But also, again with Lemma 3.7, it follows that $|C_G(a)|_p = |C_C(a)|_p |C_K(a)|_p$. Comparing these equations yields that
$$|C|_p |C_K(a)|_p |C_K(b)|_p = |G|_p = |C_G(a)|_p = |C_C(a)|_p |C_K(a)|_p$$
and hence
$$|C|_p |C_K(b)|_p = |C_C(a)|_p.$$
As $C_C(a) \leq C$, we have that $|C_C(a)|_p \leq |C|_p$ and thus $|C_K(b)|_p = 1$. This implies that $|C|_p |K|_p = |G|_p = |C|_p |C_K(a)|_p$. Therefore $|K|_p = |C_K(a)|_p$ and if $|K|_p = 1$, then C must contain a Sylow p-subgroup of G contrary to our hypothesis. So we have that $|K|_p \neq 1$ as stated. □

LEMMA 3.13. *Let $p \in \pi(G)$ and let $P \in Syl_p(G, z)$. Then $P \cap C \in Syl_p(C)$ and $|K|_p = |I_P(z)| = |P : C_P(z)|$.*

PROOF. Let $P \cap C \leq P_0 \in Syl_p(C)$. Then Lemma 3.8 yields that $P_0 \leq P_1 \in Syl_p(G, z)$ and by Lemma 3.10 there exists an element $x \in C$ such that $P = P_1^x$. But then
$$P_0^x \leq C_{P_1}(z)^x = C_P(z) = P \cap C$$
and therefore $P \cap C$ is already a Sylow p-subgroup of C. For the second statement, Lemma 3.7 gives that $|G| = |C||K|$ and thus $|P| = |G|_p = |C|_p |K|_p$. On the other hand
$$|P| = |C_P(z)||P : C_P(z)| = |C_P(z)||I_P(z)| = |C|_p |I_P(z)|$$
by the previous paragraph. Hence $|K|_p = |I_P(z)|$. □

CHAPTER 4

A Minimal Counter-Example to Glauberman's Z*-Theorem

We now begin our investigation of a minimal counter-example to the Z*-Theorem. For the remainder of this text, until stated otherwise (in the last chapter), we work under Hypothesis 4.1. The reader will be reminded of this hypothesis at various occasions, but in our results we will usually not mention it. However, there will be additional hypotheses coming in later on and these will be referred to explicitly.

HYPOTHESIS 4.1.
Let G be a counter-example to Glauberman's Z-Theorem that is minimal in the sense that*

- *if H is a proper subgroup of G, then every isolated involution of H lies in $Z^*(H)$ and*
- *if $N \trianglelefteq H \leq G$ is such that $\tilde{H} := H/N$ is a proper factor of G, then every isolated involution of $\tilde{H}$ lies in $Z^*(\tilde{H})$.*

Let z be an isolated involution in G such that $z \notin Z^*(G)$. Moreover let $C := C_G(z)$ and let M be a maximal subgroup of G containing C, let $\overline{C} := C/O(C)$ and $K := \{zz^g \mid g \in G\}$.

We note that if G is chosen to be a minimal counter-example to the Z*-Theorem with respect to the group order, then G satisfies Hypothesis 4.1. The next few lemmas capture some fundamental statements that follow from our choices in this hypothesis and that are used throughout this text many times.

LEMMA 4.2. *Suppose that $t \in G$ is an involution and suppose further that $t \in H < G$ and that t is isolated in H. Then the following hold:*

(1) $t \in Z^*(H)$, *in particular* $H = C_H(t)O(H)$.
(2) $\{tt^h \mid h \in H\} \subseteq O(H)$.
(3) $O_{2',2}(C_G(t)) \cap H \leq O_{2',2}(H)$.
(4) $\mathcal{W}_H^*(\langle t \rangle, p) \subseteq Syl_p(H)$ *for all* $p \in \pi(H)$.
(5) t *centralises* $O_2(H)E(H)$.
(6) *If* $t \notin Z(H)$, *then there exists an odd prime p such that* $[O_p(H), t] \neq 1$.

PROOF. By our hypothesis and since H is a proper subgroup of G, the Z*-Theorem holds in H. As t is isolated in H, this implies that $t \in Z^*(H)$. In particular $[H, t] \leq O(H)$ and therefore $H = C_H(t)O(H)$, giving (1). For (2) we note that $\{tt^h \mid h \in H\}$ generates $[H, t]$ which is contained in $O(H)$.

23

For (3) let $X := O_{2',2}(C_G(t)) \cap H$ and set $\widehat{H} := H/O(H)$. We note that
$$XO(C_G(t))/O(C_G(t)) \leq O_2(C_G(t)/O(C_G(t)))$$
is a 2-group whence $X/X \cap O(C_G(t))$ is a 2-group. As $t \in Z^*(H)$, we have that $t \in O_{2',2}(H)$ and consequently Lemma 2.8 is applicable. It yields that
$$X \cap O(C_G(t)) \leq O(C_H(t)) \leq O(H).$$
In particular $\widehat{X} \simeq X/X \cap O(H)$ is a 2-group. Next let $h \in C_H(t)$. Then $X^h = X$ and therefore $\widehat{X}$ is normal in $\widehat{C_H(t)}$. But $H = C_H(t)O(H)$ which means that $\widehat{H} = \widehat{C_H(t)}$ and hence $\widehat{X}$ is a normal 2-subgroup of $\widehat{H}$. It follows that $X \leq O_{2',2}(H)$.

For (4) we recall that t is isolated in H and that, therefore, Lemma 3.8 is applicable to t and H.

Next we look at (5) and consider $[E(H), t]$. By (1) we have that
$$[E(H), t] \leq E(H) \cap O(H) \leq Z(E(H))$$
because $O(H)$ is soluble. Thus $[E(H), t, E(H)] = 1$. With the Three Subgroups Lemma and since $E(H)$ is perfect it follows that $[E(H), t] = 1$. As t is isolated in H, Lemma 3.1 (2) yields that t centralises $O_2(H)$.

For (6) we assume that t centralises $F^*(H)$. Then $t \in C_H(F^*(H)) = Z(F(H))$ and therefore $t \in O_2(H)$. Then $t \in Z(H)$ by 3.1 (1), which is a contradiction. We conclude that $[F^*(H), t] \neq 1$ which, together with (5), yields that $[O(F(H)), t] \neq 1$. Therefore we find an odd prime p such that $O_p(H)$ is not centralised by t. □

LEMMA 4.3.
(1) $r_2(G) \geq 2$.
(2) G possesses at least two conjugacy classes of involutions.
(3) K generates a normal subgroup of G of even order.

PROOF. If $r_2(G) = 1$, then the Sylow 2-subgroups of G are cyclic or quaternion. In the cyclic case, Burnside's Theorem (2.19) yields that $z \in Z^*(G)$ contrary to our hypothesis. In the quaternion case, the Brauer-Suzuki Theorem (2.20) gives a similar contradiction. This proves (1). In particular it follows that C contains involutions that are distinct from z. These involutions cannot be conjugate to z which leads to (2). For (3) we note that $\langle K \rangle = [G, z] \trianglelefteq G$. If this group has odd order, then $z \in Z^*(G)$ and then G is not a counter-example. □

LEMMA 4.4. $G = F^*(G)\langle z \rangle$ and $F^*(G)$ is a non-abelian simple group.

PROOF.
(1) $O(G) = 1$ and $G = \langle z^G \rangle$.

PROOF. Suppose that $O(G) \neq 1$ and let $\widehat{G} := G/O(G)$. Lemma 3.1 (7) yields that $\widehat{z}$ is isolated in $\widehat{G}$. Then from the minimal choice of G it follows that $\widehat{z} \in Z^*(\widehat{G}) = Z(\widehat{G})$. But then $z \in Z^*(G)$, which is a contradiction.

For the second statement assume that $H := \langle z^G \rangle < G$. We note that $H \trianglelefteq G$ and hence $O(H) \leq O(G) = 1$. Together with Lemma 4.2 (1) it follows that $z \in Z^*(H) = Z(H)$ and hence z commutes with all its conjugates in G. But z is isolated and consequently $z \in Z(G)$. This is a contradiction. □

(2) $F(G) = 1$.

PROOF. Assume otherwise. Then (1) yields that $O_2(G) \neq 1$. We note that z centralises $O_2(G)$ by Lemma 3.1 (2). Therefore all conjugates of z in G centralise $O_2(G)$ and with (1) it follows that $O_2(G) \leq Z(G)$. In particular $z \notin O_2(G)$ because $z \notin Z(G)$. Now let $t \in O_2(G)$ be an involution. In the factor group $\widetilde{G} := G/\langle t \rangle$ we have that $\widetilde{z}$ is isolated by Lemma 3.1 (7). The minimal choice of G implies that $\widetilde{z} \in Z^*(\widetilde{G})$. Let $X \trianglelefteq G$ be such that $\widetilde{X} = O(\widetilde{G})$. Then $O(X) \leq O(G) = 1$ by (1) and it follows that $\langle t \rangle \in \mathrm{Syl}_2(X)$ and consequently $X = \langle t \rangle$ by Theorem 2.19. Therefore $\widetilde{z} \in Z^*(\widetilde{G}) = Z(\widetilde{G})$ which means that $z \in O_2(G)$. This is a contradiction. □

(3) $G = F^*(G)\langle z \rangle$.

PROOF. Assume that $F^*(G)\langle z \rangle < G$. Then $z \in Z^*(F^*(G)\langle z \rangle)$ by Lemma 4.2 (1) and therefore

$$[F^*(G), z] \leq F^*(G) \cap O(F^*(G)\langle z \rangle) \leq O(F^*(G)) = 1.$$

This implies that $F^*(G)$ centralises z and all its conjugates. It follows that $F^*(G) \leq Z(G) = 1$, by (1) and (2), and this is impossible. Hence $F^*(G)\langle z \rangle = G$. □

(4) $F^*(G)$ is simple and non-abelian.

PROOF. $F^*(G)$ is non-abelian by (2), in fact $F^*(G) = E(G)$. We assume that $F^*(G)$ is not simple and deduce that G has at least two components. Let N be a component of G and let $z \in S \in \mathrm{Syl}_2(C)$. Then $z \in Z(S)$ by Lemma 3.1 (2) and, as N is subnormal in G, we have that $N \cap S \in \mathrm{Syl}_2(N)$. Therefore z centralises a non-trivial 2-subgroup of N. In particular $N \cap N^z \neq 1$. But N and N^z are normal in $F^*(G)$ whence $N \cap N^z$ is normal in $F^*(G)$. As N is a minimal normal subgroup of $F^*(G)$, this implies that $N = N^z$. We recall that $N \neq F^*(G)$, so $N\langle z \rangle < G$ and Lemma 4.2 (1) yields that $z \in Z^*(N\langle z \rangle)$. It follows that z centralises every component of G whence $[F^*(G), z] = 1$, which is a contradiction.

We conclude that $F^*(G)$ is simple. □

□

LEMMA 4.5. *If $O^2(C) \neq C$ and $z \in S \in \mathrm{Syl}_2(G)$, then S is a direct product of $\langle z \rangle$ with a subgroup of S of index 2.*

PROOF. First it follows from Lemma 3.1 (2) that $z \in Z(S)$ and thus $S \leq C$. Lemma 3.1 (9) implies that Lemma 2.18 is applicable. As $O^2(C) \neq C$, it yields that $O^2(G) \neq G$ and therefore, with Lemma 4.4, that $z \notin O^2(G) = F^*(G)$. In particular we see that $S = (S \cap F^*(G)) \times \langle z \rangle$. □

LEMMA 4.6. $O^2(C_{F^*(C)}(z)) = C_{F^*(C)}(z)$ and $|C : O^2(C)| \leq 2$.
In particular $|\overline{C} : O^2(\overline{C})| \leq 2$ and $O^2(C/\langle z \rangle) = C/\langle z \rangle$.

PROOF. If G is simple, then Lemma 3.1 (10) yields the result.

Next suppose that G is not simple. Then Lemma 4.4 implies that $G = F^*(G)\langle z \rangle$ and that $O^2(G) = F^*(G)$ has index 2 in G. Let $z \in S \in \mathrm{Syl}_2(G)$ and let $T := S \cap F^*(G)$. Then $S = T \times \langle z \rangle$ by Lemma 4.5 and T is a Sylow 2-subgroup of $C_0 := C_{F^*(G)}(z)$. The simplicity of $F^*(G)$ gives that $O^2(F^*(G)) = F^*(G)$. As C_0 is z-invariant, Lemma 3.1 (9) yields that the hypothesis of Lemma 2.18 is satisfied, so we deduce that $O^2(C_0) = C_0$. As $O^2(C_0) = O^2(C) \cap C_0$, it follows that $C/O^2(C)$ has order 2 and this implies the remaining assertions. □

The next two results show that G behaves almost like a simple group.

LEMMA 4.7.
(1) $F^*(G)$ is the unique minimal normal subgroup of G.
(2) Let $t \in G$ be an isolated involution in G. Then $G = F^*(G)\langle t \rangle$.
(3) $G = \langle K, z \rangle$.

PROOF. Lemma 4.4 implies (1). For (2) we see that t is isolated and hence Lemma 4.4 may be applied to G and t instead of G and z. We know that $\langle K \rangle \trianglelefteq G$ by Lemma 4.3 (3) and hence (1) forces $F^*(G) \leq \langle K \rangle$. This yields (3). □

COROLLARY 4.8. *Suppose that $t \in G$ is an isolated involution and that H is a maximal subgroup of G containing t. Then H is primitive.*

PROOF. Let $1 \neq X \trianglelefteq H$. Then the maximality of H implies that $N_G(X) = G$ or $N_G(X) = H$. In the first case, it follows from Lemma 4.7 that $G = X\langle t \rangle \leq H$, which is a contradiction. Therefore the second case holds and consequently H is primitive. □

The last few lemmas of this section are taken from [**Wal08**], with minimal changes.

LEMMA 4.9. *Suppose that $t \in z^G \setminus M$ and set $n := |M : C|$. Let $D := M \cap M^t$ and let $I := I_D(t)$. Then the following hold:*
(1) $D = O(D)C_D(t)$.
(2) $z^M = \{z^x \mid x \in I\}$. *In particular D is transitive on z^M.*
(3) $M = CI$. *More precisely, every coset of C in M contains exactly one element of I.*
(4) $|I| = |D : C_D(t)| = |D : C_D(z)| = n$.
(5) *Let $q \in \pi(G)$ and $Q \in \mathrm{Syl}_q(D, t)$. Then $|I_Q(t)| = n_q$.*

PROOF. As $\langle D, z \rangle \leq M \neq G$, Lemma 4.7 yields that D is not normal in G. Moreover D is t-invariant and therefore $D\langle t \rangle$ is a proper subgroup of G. Since t is isolated in G, it follows from Lemma 4.2 (1) that
$$D\langle t \rangle = O(D\langle t \rangle)C_{D\langle t \rangle}(t).$$
Hence $[D, t] \leq D \cap O(D\langle t \rangle) \leq O(D)$ which gives (1).

Let $u \in z^M$. Then u and z^t are conjugate in $\langle u, z^t \rangle$ because uz^t has odd order by Lemma 3.1 (3). In fact there exists an involution $s \in \langle u, z^t \rangle$ such that $u^s = z^t$. Now $u = z^{ts}$. On the other hand, since $u \in z^M$, Lemma 3.1 (4) yields that z and u

are also conjugate in M. Choose $x \in M$ such that $u^x = z$. Then $z = u^x = z^{tsx}$ and therefore $tsx \in C$. This yields that $ts \in M$ because $\langle x, C \rangle \leq M$. As ts is inverted by t, it follows that $ts \in M \cap M^t = D$ and thus $ts \in I$. This gives (2) and implies that $M = CI$. To finish the proof of (3), let $x_1, x_2 \in I$ be such that $Cx_1 = Cx_2$. Then $x_1 x_2^{-1} \in C$. But $x_1 t$ and $x_2 t$ are involutions that are conjugate to t and hence to z. Therefore $x_1 t x_2 t = x_1 t (x_2 t)^{-1} = x_1 x_2^{-1} \in C$. Lemma 3.1 (11) implies that $x_1 t = x_2 t$ and finally $x_1 = x_2$.

For (4), we apply Lemma 3.7 to the isolated involution t in $D\langle t \rangle$ and it follows that I is a set of representatives for the cosets of $C_D(t)$ in D. To prove (5) we observe that, since t is isolated in $D\langle t \rangle$, we may apply Lemma 3.13. From there we obtain that $C_Q(t) \in \mathrm{Syl}_q(C_D(t))$ and that $n_q = |D : C_D(t)|_q = |Q : C_Q(t)| = |I_Q(t)|$. □

LEMMA 4.10. *Suppose that C is a maximal subgroup of G and let $p \in \pi(F(C))$. Then C contains a Sylow p-subgroup of G and every z-invariant p-subgroup of G is centralised by z.*

PROOF. Let $P \in \mathrm{Syl}_p(C)$. Then $z \in C_G(P) \leq C_G(O_p(C))$. But C is maximal in G and therefore Corollary 4.8 implies that $N_G(O_p(C)) = C$. It follows that $C_G(P) \leq C$. Now if we set $X := C_G(P)$ and $Y := N_G(P)$, then Lemma 3.1 (6) yields that $Y = XC_Y(z)$. But X and $C_Y(z)$ are both contained in C, thus $N_G(P) = Y \leq C$. This means that $P \in \mathrm{Syl}_p(G)$. The rest follows from Lemmas 3.8 and 3.10. □

LEMMA 4.11. *Suppose that C is a maximal subgroup of G and let $\pi := \pi(F(C))$. Let $z \in H < G$. Then $[H, z]$ is a π'-group.*

PROOF. Let $H_0 := [H, z]$ and assume that $p \in \pi \cap \pi(H_0)$. From Lemma 4.2 (2) we know that H_0 is of odd order and that z acts coprimely on H_0. Then Lemma 2.1 (7) implies that $\mathcal{W}_{H_0}^*(\langle z \rangle, p) \subseteq \mathrm{Syl}_p(H_0)$. Lemma 4.10 yields that every z-invariant p-subgroup of H_0 is centralised by z. Then it follows with Lemma 2.5 that $H_0 = C_{H_0}(z) O_{p'}(H_0)$. But this means that $H_0 = [H_0, z] \leq O_{p'}(H_0)$, contrary to our choice of p. □

LEMMA 4.12. *Suppose that q is a prime such that $O_q(M) \not\leq C$. Then M does not contain a Sylow q-subgroup of G.*

PROOF. First we observe that q is odd by Lemma 3.1 (2). With Lemma 4.2 (4) we choose $Q \in \mathrm{Syl}_q(M, z)$ and assume that $Q \in \mathrm{Syl}_q(G, z)$. As $O_q(M) \not\leq C$, we have that $1 \neq X := I_{O_q(M)}(z)$. If we set $n := |M : C|$, then Lemma 3.13 implies that $1 \neq |I_Q(z)| = n_q$. Our objective is to show that X lies in every conjugate of M in G.

We see that X is C-invariant and hence Lemma 3.10 gives that X is contained in every z-invariant Sylow q-subgroup of G. The same lemma and our assumption that $Q \in \mathrm{Syl}_q(G, z)$ imply that every z-invariant q-subgroup of G lies in M. Now let $g \in G \backslash M$ and $M_1 := M^g$. We look at $D := M_1 \cap M_1^z$ and see that $D = C_D(z)O(D)$ and $|D : C_D(z)| = n$ by Lemma 4.9 (1) and (4). If we choose $T \in \mathrm{Syl}_q(D, z)$, then part (5) of the same lemma yields that $|I_T(z)| = n_q \neq 1$. Moreover $T \leq M$ because

T is z-invariant. Then there exists an element $c \in C$ such that $I_T(z) = I_{Q^c}(z) = (I_Q(z))^c$ and finally $X = X^{c^{-1}} \subseteq I_T(z) \subseteq D \leq M_1$. Hence

$$1 \neq X \subseteq N := \bigcap_{g \in G} M^g \trianglelefteq G.$$

But we know from Corollary 4.8 that M does not contain any non-trivial normal subgroup of G, so this is impossible. $\square$

LEMMA 4.13. *Let $t \in C \backslash \{z\}$ be an involution. Then $C_G(t) \not\leq M$.*

PROOF. Let $w := zt$. Then we have that $1 \neq \langle w^G \rangle \trianglelefteq G$ and hence $\langle w^G \rangle \not\leq M$ by Corollary 4.8. So there exists a conjugate u of w that is not contained in M and thus does not centralise z. We note that w and z are distinct and commute. As z is isolated, this implies that w and z are not conjugate and it follows that u and z are not conjugate. Now set $D := \langle u, z \rangle$. By Lemma 3.1 (5) we know that the order of uz is even and not divisible by 4. More precisely the Sylow 2-subgroups of D are elementary abelian of order 4 and contain the unique central involution v of D. As $u \in C_G(v)$ and $u \notin M$, we have that $C_G(v) \not\leq M$. Let $z \in T \in \mathrm{Syl}_2(D)$ and let $d \in D$ be such that $u^d \in T$. It follows that $T = \langle z, u^d \rangle$ and hence $v = zu^d$. But $u^d \ (= zv)$ and $w \ (= zt)$ both centralise z and therefore they are conjugate in C by Lemma 3.1 (9). Thus $v = zu^d$ and $t = zw$ are conjugate in C and $C_G(v) \not\leq M$ implies that $C_G(t) \not\leq M$. $\square$

CHAPTER 5

Balance and Signalizer Functors

The concept of signalizer functors was introduced by Gorenstein and has many applications in finite group theory, for example it plays an important role in the Classification of Finite Simple Groups. Since we keep working under Hypothesis 4.1, the special behaviour of z in our minimal counter-example G leads to signalizer functors for the prime 2 quite naturally and it turns out that they become powerful tools. In this section, we therefore recall the notion of a signalizer functor and Glauberman's Soluble Signalizer Functor Theorem. Moreover we introduce particular balance conditions and special signalizer functors that will have a role to play in our analysis later on.

DEFINITION 5.1. Suppose that A is an elementary abelian 2-subgroup of C that contains z.

- For all $a \in A^\#$, we set

$$\alpha(a) := O(C_C(a)),$$
$$\gamma(a) := [C_G(a), z]C_{O(C)}(a) \text{ and}$$
$$\Theta(a) := \gamma(a)O(C_C(A)).$$

- Suppose that A is of rank at least 3. We say that A is **balanced** if and only if for all $a \in A^\#$ we have that $\alpha(a) \leq O(C)$. We say that A is **weakly balanced** if and only if for all $a \in A^\#$ we have that $\alpha(a) \leq O(C_C(A))O(C)$.
- For all subgroups V of A of order 4 we set

$$\Delta_V := \bigcap_{v \in V^\#} O(C_G(v)) \cap C.$$

We say that A is **2-balanced** if A is of rank at least 4 and if for all subgroups V of A of order 4 we have that $\Delta_V \leq O(C)$.

The last definition seems to be slightly different from Gorenstein's notion of "2-balance for the prime 2" in [**Gor82**], Section 4.4, but it is in fact the same because for all involutions $a \in C$ we know that $C_G(a) < G$ by Lemma 4.7 and hence $C_G(a) = C_C(a)O(C_G(a))$ by Lemma 4.2 (1). Also, the words "balanced" and "weakly balanced" have a meaning in standard literature already (see for example [**GLS96**]), but the definitions above are close enough to these standard notions that we thought that the choice of words is appropriate.

DEFINITION 5.2. Let p be a prime and let A be an abelian p-subgroup of G. Suppose that for all $a \in A^{\#}$ there is defined an A-invariant p'-subgroup $\theta(a)$ of $C_G(a)$. Then θ is an **A-signalizer functor** if and only if the following balance condition is satisfied for all $a, b \in A^{\#}$:

$$\theta(a) \cap C_G(b) \leq \theta(b).$$

θ is **soluble** if and only if $\theta(a)$ is soluble for all $a \in A^{\#}$. A (soluble) A-signalizer functor θ is said to be **complete** if and only if there exists a (soluble) A-invariant p'-subgroup W of G such that $\theta(a) = C_W(a)$ for all $a \in A^{\#}$. We refer to such a subgroup W as the **completion** for θ.

THEOREM 5.3 (Glauberman's Soluble Signalizer Functor Theorem). *Suppose that p is a prime and that $A \leq G$ is an abelian p-subgroup of rank at least 3.*

If θ is a soluble A-signalizer functor, then θ is complete. In particular, the completion $\langle \theta(a) \mid a \in A^{\#} \rangle$ is a soluble p'-subgroup of G.

PROOF. This holds independently of Hypothesis 4.1, see [**Gla76**]. □

REMARK 5.4. It follows from the Odd Order Theorem that signalizer functors for 2-groups are always soluble.

LEMMA 5.5. *Suppose that A is an elementary abelian 2-subgroup of C that contains z.*
  (1) *For all $a \in A^{\#}$, we have that $\alpha(a) \leq O(C_G(a))$ and $\alpha(a) = \alpha(az)$.*
  (2) *If A is balanced, then γ defines an A-signalizer functor.*
  (3) *If A is weakly balanced, then Θ defines an A-signalizer functor.*

PROOF.
  (1) Let $a \in A^{\#}$. We know that $z \in C_G(a)$, so Lemma 4.7 implies that $C_G(a) < G$ and then $z \in Z^*(C_G(a))$ with Lemma 4.2 (1). In particular $z \in O_{2',2}(C_G(a))$ and $[C_G(a), z] \leq O(C_G(a))$. Now Lemma 2.8 yields that

$$\alpha(a) = O(C_{C_G(a)}(z)) \leq O(C_G(a)).$$

The fact that $C_C(a) = C_C(az)$ implies the next statement.

  (2) Suppose that A is balanced and let $a, b \in A^{\#}$. As $C_{O(C)}(a)$ normalises $[C_G(a), z]$ and these two subgroups are A-invariant $2'$-subgroups of $C_G(a)$, it follows that $\gamma(a)$ is an A-invariant $2'$-subgroup of $C_G(a)$. Now we apply Lemma 2.1 (3) to deduce that

$$\gamma(a) \cap C_G(b) = C_{[C_G(a),z]}(b)(C_{O(C)}(a) \cap C_G(b)),$$

where the second factor lies in $C_{O(C)}(b)$ and hence in $\gamma(b)$.

Let $X := C_{[C_G(a),z]}(b)$. Then the coprime action of z on X yields, with Lemma 2.1 (2), that $X = C_X(z)[X, z]$. The commutator $[X, z]$ is contained in $[C_G(b), z]$ and hence in $\gamma(b)$. So it is left to show that $C_X(z) \leq \gamma(b)$ to establish the balance condition. But

$$C_X(z) = [C_G(a), z] \cap C_C(b) \leq O(C_G(a)) \cap C_C(b)$$
$$\leq O(C_C(a)) \cap C_G(b) = \alpha(a) \cap C_G(b) \leq C_{O(C)}(b) \leq \gamma(b)$$

because A is balanced. Hence γ defines an A-signalizer functor.

(3) Suppose that A is weakly balanced. For all $a \in A^{\#}$, we see as in (2) that $\Theta(a)$ is an A-invariant $2'$-subgroup of $C_G(a)$. For the balance condition let $a, b \in A^{\#}$. Two applications of Lemma 2.1 (3) give that

$$\Theta(a) \cap C_G(b)$$
$$= C_{[C_G(a),z]}(b) \cdot (C_{O(C)}(a) \cap C_G(b)) \cdot (O(C_C(A)) \cap C_G(b))$$
$$\leq C_{[C_G(a),z]}(b) C_{O(C)}(b) O(C_C(A)).$$

The second and the third factor are contained in $\Theta(b)$, so let $Y := C_{[C_G(a),z]}(b)$. Then

$$C_Y(z) \leq C_G(b) \cap O(C_G(a)) \cap C \leq C_G(b) \cap \alpha(a).$$

As A is weakly balanced, Lemma 2.1 (3) implies that

$$C_Y(z) \leq C_{\alpha(a)}(b) \leq C_{O(C)}(b) C_{O(C_C(A))}(b) = C_{O(C)}(b) O(C_C(A)) \leq \Theta(b)$$

But also $[Y, z] \leq [C_G(b), z] \leq \Theta(b)$, so with the coprime action of z on Y and Lemma 2.1 (2) it follows that $Y = C_Y(z)[Y, z] \leq \Theta(b)$. $\square$

In many situations, we use specifically designed signalizer functors to show that C cannot have large elementary abelian 2-subgroups (or at least that we can control where they lie in C). The idea is to find a suitable signalizer functor that is complete and such that its completion, a $2'$-subgroup of G, contains the set K. This contradicts Lemma 4.3 (3). This argument is captured more specifically in the next lemma and we see some applications in the course of this section.

LEMMA 5.6. *Suppose that A is an elementary abelian 2-subgroup of G containing z and suppose that θ is a soluble A-signalizer functor of G. Suppose further, for all $a \in A^{\#}$, that $[C_G(a), z] \leq \theta(a)$. Then A is of rank at most 2.*

PROOF. Assume that A has rank 3 or more. Then Theorem 5.3 implies that θ is complete. In particular, the completeness subgroup $W := \langle \theta(a) \mid a \in A^{\#} \rangle$ is a subgroup of G of odd order. Now let $a \in A^{\#}$ and $a \neq z$. Then $z \in C_G(a)$ and therefore, by definition of the set K, we have that $C_K(a) \subseteq [C_G(a), z] \leq \theta(a)$. Theorem 3.6 forces

$$K \subseteq \langle C_K(a), C_K(az) \rangle \leq \langle \theta(a), \theta(az) \rangle \leq W.$$

Hence K generates a subgroup of G of odd order contradicting Lemma 4.3 (3). $\square$

LEMMA 5.7. *C does not possess any 2-balanced subgroups.*

PROOF. Assume that $A \leq C$ is a 2-balanced subgroup. Then $z \in A$ and $r(A) \geq 4$ by definition, so we are aiming for a contradiction to Lemma 5.6 by showing that γ defines an A-signalizer functor. Hence let $a, b \in A^{\#}$. As b acts coprimely on $\gamma(a)$, we have that

$$\gamma(a) \cap C_G(b) = C_{[C_G(a),z]}(b) C_{C_{O(C)}(a)}(b)$$

by Lemma 2.1 (3). But we see that $C_G(b) \cap C_{O(C)}(a) \leq C_{O(C)}(b) \leq \gamma(b)$ at once. Now if we let $H := [C_G(a), z]$ and $H_0 := C_H(b)$, then it is only left to show that H_0 is contained in $\gamma(b)$.

First the coprime action of z on H_0 yields that $H_0 = C_{H_0}(z)[H_0, z]$ by Lemma 2.1 (2). Furthermore we have that $[H_0, z] \leq [C_G(b), z] \leq \gamma(b)$. Our objective is to apply Theorem 2.7. First we observe that $z \in H < G$ by Lemma 4.7 and hence $H \leq O(C_G(a))$ by Lemma 4.2 (1). In particular H has odd order and $H = [H, z]$. With $\langle z \rangle$, H and $\langle b \rangle$ playing the roles of A_0, X and B in Theorem 2.7, respectively, and with $Hyp^2(A)$ denoting the set of subgroups of A of index 4, we obtain that

$$C_{H_0}(z) = C_H(z) \cap C_G(b) = \langle [C_H(Y), z] \cap C_{H_0}(z) | Y \in Hyp^2(A) \rangle.$$

Let $Y \in Hyp^2(A)$. By hypothesis, the rank of A is at least 4 and therefore Y contains a subgroup V of order 4. With the notation from Definition 5.1 we see that $\Delta_V \leq O(C)$ because of our assumption that A is 2-balanced.

Since $z \in C_G(V) < G$, we know that $[C_G(v), z] \leq O(C_G(v))$ for all $v \in V^\#$, by Lemma 4.2 (1). This yields, with 2-balance:

$$[C_H(Y), z] \cap C_{H_0}(z) \leq [C_G(V), z] \cap C_{H_0}(z) \leq \bigcap_{v \in V^\#} [C_G(v), z] \cap C_{H_0}(z)$$

$$\leq \bigcap_{v \in V^\#} O(C_G(v)) \cap C_{H_0}(z) \leq \Delta_V \cap C_{H_0}(z) \leq O(C) \cap C_{H_0}(z) \leq C_{O(C)}(b) \leq \gamma(b).$$

Finally $C_{H_0}(z) \leq \gamma(b)$ as required and Lemma 5.6 gives a contradiction. □

We want to point out that the signalizer functor γ used in Lemma 5.7 appears in Section 4.4 of [**Gor82**], but there it is established in a different way. We decided to give an alternative approach here. The impact of Lemma 5.7 on the structure of $F^*(\overline{C})$ will be discussed in detail in Chapter 10. Now we look at the other concepts of balance. We note that, in the definition of γ, the fact that $\gamma(a)$ contains $[C_G(a), z]$ for all $a \in A^\#$ together with Lemma 5.6 implies that A has rank at most 2 or that γ is not allowed to be a signalizer functor. A similar statement holds for Θ. As this idea will be referred to quite frequently, it is convenient to have it captured in a result to be quoted later on.

LEMMA 5.8. *C does not contain any balanced or weakly balanced 2-subgroups.*

PROOF. Assume otherwise and let $A \leq C$ be an elementary abelian 2-subgroup such that A is (weakly) balanced. Then the definition of (weakly) balanced subgroups yields that $z \in A$ and $r(A) \geq 3$. Now Lemma 5.5 (2) or (3), respectively, imply that γ or (in the weakly balanced case) Θ defines an A-signalizer functor, contrary to Lemma 5.6. □

The notion of core-separated subgroups appears in [**Gol75**]. We work with a variation of it in our specific context and give an explicit proof of Goldschmidt's result that, under certain conditions, core-separated subgroups lead to a signalizer functor.

DEFINITION 5.9. Suppose that A_1 and A_2 are distinct commuting elementary abelian subgroups of G such that $A_1 \cap A_2 = 1$. Set $A := A_1 \times A_2$. Then A_1, A_2 are said to be **core-separated** if and only if for all $a \in A^\#$ and for every 2-component E of $C_G(a)$ we have that

$$[E, A_1] \leq O(C_G(a)) \text{ or } [E, A_2] \leq O(C_G(a)).$$

Before we state and prove Goldschmidt's theorem, we present two of his results from [**Gol75**], one of them much simplified. They play a role in the proof, but nowhere else in this text and they do not depend on Hypothesis 4.1.

LEMMA 5.10. *Suppose that H is a proper subgroup of G with $O(H) = 1$ and let $t \in G$ be an involution. If Y is a non-trivial subgroup of $O(C_H(t))$, then H possesses a $\langle t, Y \rangle$-invariant component that is centralised by neither t nor Y.*

PROOF. This is (2.6) in [**Gol75**]. □

LEMMA 5.11. *Suppose that A is an abelian 2-subgroup of G and that $B \leq A$. If Y is an A-invariant $2'$-subgroup of G and $Y = [Y, A]$, then*
$$C_Y(B) = \langle C_Y(B) \cap [C_Y(B_0), A] \mid B/B_0 \text{ is cyclic} \rangle.$$

PROOF. This is a very special case of (2.5) in [**Gol75**]. □

LEMMA 5.12. *Suppose that $A_1, A_2 \leq G$ are core-separated subgroups of G of order 4 and let $A := A_1 \times A_2$. For all $a \in A^\#$ set*
$$\theta(a) := \bigcap_{i=1,2} [O(C_G(a)), A_i](O(C_G(a)) \cap O(C_G(A_i))).$$
Then θ defines a soluble A-signalizer functor.

PROOF. We follow Goldschmidt's arguments in [**Gol75**]. For all $a \in A^\#$ and $i \in \{1,2\}$, we set
$$\gamma_i(a) := [O(C_G(a)), A_i](O(C_G(a)) \cap O(C_G(A_i))).$$
Let $a \in A^\#$ and $i \in \{1,2\}$. Set $X := [O(C_G(a)), A_i]$ and let $b \in A_i^\#$.

(1) $Y := [C_X(b), A_i]$ is contained in $O(C_G(b))$.

PROOF. Set $H := C_G(b)$ and $\widehat{H} := H/O(H)$. First we note that $a \in H$ because A is abelian. Moreover $Y \leq O(C_G(a))$ and hence $\widehat{Y} \leq O(C_{\widehat{H}}(\widehat{a}))$.

Let us assume that $\widehat{Y} \neq 1$. Then Lemma 5.10 yields a 2-component E of H such that $\widehat{E}$ is normalised, but not centralised by $\widehat{Y}$ and by $\widehat{a}$. Let $A_j := A_{3-i}$. Then it follows from the fact that A_1 and A_2 are core-separated that $\widehat{E}$ commutes with $\widehat{A_i}$ or with $\widehat{A_j}$.

If $[\widehat{E}, \widehat{A_i}] = 1$, then $[\widehat{A_i}, \widehat{E}, \widehat{Y}] = 1 = [\widehat{E}, \widehat{Y}, \widehat{A_i}]$ and therefore $[\widehat{Y}, \widehat{E}] = [\widehat{Y}, \widehat{A_i}, \widehat{E}] = 1$ by definition of $\widehat{Y}$ and by the Three Subgroups Lemma. This is impossible and thus $[\widehat{E}, \widehat{A_j}] = 1$. Let $a_i \in A_i$ and $a_j \in A_j$ be such that $a = a_i a_j$. Then $[\widehat{E}, \widehat{a_j}] \leq [\widehat{E}, \widehat{A_j}] = 1$, but we noted above that $[\widehat{E}, \widehat{a}] \neq 1$ and therefore $[\widehat{E}, \widehat{a_i}] \neq 1$. At the same time, we know that $[\widehat{E}, \widehat{b}] = 1$ and in particular $b \neq a_i$. We conclude that $A_i = \langle a_i, b \rangle$ because by hypothesis $|A_i| = 4$. Since Y centralises a and b, it follows that
$$\widehat{Y} = [\widehat{Y}, \widehat{A_i}] = [\widehat{Y}, \widehat{a_i}] = [\widehat{Y}, \widehat{a_j}].$$

Thus $[\widehat{Y}, \widehat{E}] = [\widehat{Y}, \widehat{a}_j, \widehat{E}] = 1$ by the Three Subgroups Lemma, which is a contradiction. Hence we have that $\widehat{Y} = 1$ as stated. □

(2) $\gamma_i(a) \cap C_G(A_i) \leq O(C_G(A_i))$.

PROOF. By definition of $\gamma_i(a)$ and Lemma 2.1 (3), we have that
$$C_{\gamma_i(a)}(A_i) = ([O(C_G(a)), A_i] \cap C_G(A_i)) \cdot (O(C_G(a)) \cap O(C_G(A_i))).$$
We see that the second factor is contained in $O(C_G(A_i))$. Therefore it suffices to prove that
$$C_X(A_i) = [O(C_G(a)), A_i] \cap C_G(A_i) \leq O(C_G(A_i)),$$
with our notation introduced before (1). Lemma 5.11, applied to A_i and to the A_i-invariant $2'$-subgroup X, yields that
$$C_X(A_i) = \langle C_X(A_i) \cap [C_X(b), A_i] \mid b \in A_i^{\#} \rangle.$$
We proved in (1) that, for all $b \in A_i^{\#}$, the subgroup $[C_X(b), A_i]$ is contained in $O(C_G(b))$. As $C_G(A_i) \leq C_G(b)$, it follows that $C_X(A_i) \leq O(C_G(A_i))$. □

(3) Let $a, b \in A^{\#}$ be arbitrary. Then $\theta(a) \cap C_G(b) \leq O(C_G(b))$.

PROOF. Let $D := \theta(a) \cap C_G(b)$, let $H := C_G(b)$ and $\widehat{H} := H/O(H)$ and assume that $\widehat{D} \neq 1$. By definition of θ, we know that $\theta(a) \leq O(C_G(a))$ and hence $\widehat{D} \leq O(C_{\widehat{H}}(\widehat{a}))$. Then Lemma 5.10 gives us a component $\widehat{E}$ of $\widehat{H}$ that is normalised, but not centralised by $\widehat{D}$ and normalised, but not centralised by $\widehat{a}$. As A_1 and A_2 are core-separated, we know that $\widehat{E}$ centralises $\widehat{A_1}$ or $\widehat{A_2}$. By symmetry we may suppose that $[\widehat{E}, \widehat{A_1}] = 1$. Then Lemma 2.1 (2) implies that $\widehat{D} = [\widehat{D}, \widehat{A_1}] C_{\widehat{D}}(\widehat{A_1})$ and the Three Subgroup Lemma yields that $[\widehat{D}, \widehat{A_1}, \widehat{E}] = 1$. Also, we know from (2) that
$$C_D(A_1) \leq \gamma_1(a) \cap C_G(A_1) \leq O(C_G(A_1))$$
whence $C_{\widehat{D}}(\widehat{A_1}) \leq O(C_{\widehat{H}}(\widehat{A_1}))$. But $\widehat{E}$ is a component of $C_{\widehat{H}}(\widehat{A_1})$ and therefore $[C_{\widehat{D}}(\widehat{A_1}), \widehat{E}] \leq [O(C_{\widehat{H}}(\widehat{A_1})), \widehat{E}] = 1$. Consequently $[\widehat{D}, \widehat{E}] = 1$, which is a contradiction. We deduce that $\widehat{D} = 1$, so $\theta(a) \cap C_G(b) \leq O(C_G(b))$ as stated. □

(4) θ defines a signalizer functor.

PROOF. We only need to establish the balance condition. Hence let $a, b \in A^{\#}$, let $i \in \{1, 2\}$ and set $X := \theta(a) \cap C_G(b)$. From (3) we know that $X \leq O(C_G(b))$ and therefore
$$[X, A_i] \leq [O(C_G(b)), A_i] \leq \gamma_i(b).$$
Moreover (2) yields that $C_X(A_i) \leq C_{\gamma_i(a)}(A_i) \leq O(C_G(A_i))$ whence
$$C_X(A_i) \leq O(C_G(A_i)) \cap O(C_G(b)) \leq \gamma_i(b).$$
Hence $[X, A_i]$ and $C_X(A_i)$ are contained in $\gamma_i(b)$. Lemma 2.1 (2) gives that $X = [X, A_i] C_X(A_i)$ which implies that $X \leq \gamma_1(b) \cap \gamma_2(b) = \theta(b)$. □

This completes the proof. □

LEMMA 5.13. *C does not possess any core-separated elementary abelian subgroups A_1, A_2 of order 4.*

PROOF. Assume that such a pair of core-separated subgroups exists. Let $A := A_1 \times A_2$ and consider Goldschmidt's signalizer functor, i.e. for all $a \in A^\#$ define

$$\theta(a) := \bigcap_{i=1,2} [O(C_G(a)), A_i](O(C_G(a)) \cap O(C_G(A_i))),$$

as in Lemma 5.12. This is a soluble A-signalizer functor of G. Now let $a \in A^\#$, let $H := C_G(a)$ and $i \in \{1, 2\}$. As z lies in H and in $C_G(A_i)$, we first have that H and $C_G(A_i)$ are proper subgroups of G, by Lemma 4.7. Then we deduce that $[H, z] \leq O(H)$ and $[C_G(A_i), z] \leq O(C_G(A_i))$ with Lemma 4.2 (2). Let $H_0 := [H, z]$. Then Lemma 2.1 (2) implies that

$$H_0 = [H_0, z] = [C_{H_0}(A_i)[H_0, A_i], z] = \langle [C_{H_0}(A_i), z], [H_0, A_i] \rangle.$$

Now $[C_{H_0}(A_i), z] \leq O(H) \cap O(C_G(A_i))$ and $[H_0, A_i] \leq [O(H), A_i]$, consequently $H_0 \leq \theta(a)$.

So we have established that $[C_G(a), z] \leq \theta(a)$ for all $a \in A^\#$. But A has rank 4 and therefore this contradicts Lemma 5.6. □

We include here another simple application of our signalizer functor results, again following an idea of Goldschmidt's, but this time from [**Gol72**]. After that we finish this section with a technical lemma that is going to be applied from Chapter 10 onwards.

LEMMA 5.14. *Suppose that $V \leq O_{2',2}(C)$ is an elementary abelian subgroup of order 4 containing z. Then $r_2(C_G(V)) = 2$.*

PROOF. Let us assume that this is false and let A be an elementary abelian subgroup of $C_G(V)$ of order 8 containing V. For all involutions $a \in A$ we have that $z \in C_G(a) < G$ by Lemma 4.7 and therefore $V \leq O_{2',2}(C) \cap C_G(a) \leq O_{2',2}(C_G(a))$ by Lemma 4.2 (3). For all $a \in A^\#$ we set

$$\delta(a) := [O(C_G(a)), V](O(C_G(V)) \cap C_G(a)).$$

(1) Let $a, b \in A^\#$ and $X := \delta(a) \cap C_G(b)$. Then $[X, V] \leq \delta(b)$.

PROOF. Set $H := C_G(b)$ and recall that $V \leq O_{2',2}(H)$. As X has odd order, we obtain that $[X, V] \leq X \cap O_{2',2}(H) \leq O(H)$. Then Lemma 2.1 (2) gives that $[X, V] = [X, V, V] \leq [O(H), V] \leq \delta(b)$. □

(2) Let $a, b \in A^\#$ and $X := \delta(a) \cap C_G(b)$. Then $C_X(V) \leq \delta(b)$.

PROOF. We show that $C_{\delta(a)}(V) \leq O(C_G(V))$ because this implies that $C_X(V) \leq O(C_G(V)) \cap C_G(b) \leq \delta(b)$ as desired. By Lemma 2.1 (3) we have that

$$C_{\delta(a)}(V) = C_{[O(C_G(a)), V]}(V)(O(C_G(V)) \cap C_G(a))$$

and the second factor is contained in $O(C_G(V))$. So we need to prove that $C_{[O(C_G(a)), V]}(V) \leq O(C_G(V))$. Set $Y := [O(C_G(a)), V]$. Theorem

2.7, applied with 1, A, Y and V in the roles of B, A, X and A_0 yields that
$$C_Y(V) = \langle [C_Y(t), V] \cap C_Y(V) \mid t \in A^\# \rangle.$$
Hence let $t \in A^\#$ and $Y_0 := [C_Y(t), V] \cap C_Y(V)$. It suffices to show that $Y_0 \leq O(C_G(V))$. As $V \leq O_{2',2}(C_G(t))$ by Lemma 4.2 (3), we have that $Y_0 \leq O(C_G(t))$. Moreover $Y_0 \leq O(C_G(a))$, so if t or a is contained in V, then $Y_0 \leq O(C_G(V))$. Thus we suppose that neither a nor t is contained in V. Then there exists some $v \in V$ such that $t = av$ (because $|A| = 8$ and $|V| = 4$) and so $Y_0 \leq C_Y(V) \leq C_G(v)$. Again by Lemma 4.2 (3), we know that $V \leq O_{2',2}(C_G(v))$ and hence $Y_0 \leq O(C_G(v))$. Therefore $Y_0 \leq O(C_G(V))$, completing the proof. □

(3) δ defines a soluble A-signalizer functor.

PROOF. Let $a, b \in B^\#$. We see that $\delta(a)$ is an A-invariant $2'$-subgroup (hence a soluble subgroup) of $C_G(a)$ and so we only need to establish the balance condition. But the work is already done: Lemma 2.1 (2) yields that $X = [X, V]C_X(V)$ and we know from (1) and (2) that $[X, V]$ and $C_X(V)$ are contained in $\delta(b)$. □

As δ defines a soluble A-signalizer functor and as, for all $a \in A^\#$, we have that
$$[C_G(a), z] \leq [O(C_G(a)), z] \leq [O(C_G(a)), V] \leq \delta(a),$$
so that Lemma 5.6 is applicable. It yields a contradiction. □

LEMMA 5.15. *Suppose that $E \in \mathcal{L}_2(C)$ and that $a \in E$ is an involution. Then $\overline{\alpha(a)}$ centralises $O_2(\overline{C})$ as well as every component of $\overline{C}$ distinct from $\overline{E}$. If $O(C_E(a)) \leq O(C)$ and $\alpha(a) \not\leq O(C)$, then $\overline{\alpha(a)}$ induces non-trivial outer automorphisms on $\overline{E}$ of odd order.*

PROOF. Let $a \in E$ be an involution. The result is immediate if $a = z$, so we suppose that $a \neq z$. As $O_2(\overline{C}) \leq C_{\overline{C}}(\overline{a})$, it follows that
$$[\overline{\alpha(a)}, O_2(\overline{C})] \leq [O(C_{\overline{C}}(\overline{a})), O_2(C_{\overline{C}}(\overline{a}))] = 1.$$
Also, for all components $\overline{L}$ of $\overline{C}$ distinct from $\overline{E}$, we see that $\overline{L} \leq E(C_{\overline{C}}(\overline{a}))$ and thus
$$[\overline{\alpha(a)}, \overline{L}] \leq [O(C_{\overline{C}}(\overline{a})), E(C_{\overline{C}}(\overline{a}))] = 1.$$
Now we suppose that $O(C_E(a)) \leq O(C)$ and $\alpha(a) \not\leq O(C)$. Hence there exists some $x \in C$ of odd order such that $1 \neq \overline{x} \in O(C_{\overline{C}}(\overline{a}))$. Then $\overline{x} \notin Z(F^*(\overline{C}))$ because this is a 2-group. The first part of the lemma yields that $\overline{x}$ centralises $O_2(\overline{C})$ and every component of $\overline{C}$ distinct from $\overline{E}$ (if any exists). Therefore $\overline{x}$ does not centralise $\overline{E}$. If $\overline{x}$ induces an inner automorphism on $\overline{E}$, then $\overline{x} \in \overline{E}C_{\overline{C}}(\overline{E})$. It follows that $\overline{x} \in O(C_{\overline{E}}(\overline{a}))O(C_{\overline{C}}(\overline{E})) = 1$, with our hypothesis that $O(C_E(a)) \leq O(C)$. But this is impossible. Therefore $\overline{x}$ induces a non-trivial outer automorphism of $\overline{E}$ of odd order. □

CHAPTER 6

Preparatory Results for the Local Analysis

Here we introduce the Bender Method, one of our most important tools for local arguments. It comes into action for the first time when we analyse the behaviour of isolated involutions in proper subgroups of G. The results that we obtain are very often applied to maximal subgroups of G containing the centraliser of an involution, and therefore they play a role not only in the proof of Theorem **A**, but also in Chapters 8, 13 and 14. Throughout, we assume Hypothesis 4.1. Some of the material in this section is taken from [**Wal09**].

6.1. The Bender Method

DEFINITION 6.1. Let H_1 and H_2 be maximal subgroups of G. Then we say that H_1 **infects** H_2 and we write $H_1 \hookrightarrow H_2$ if and only if there exists a subgroup A of $F(H_1)$ such that $AC_{F^*(H_1)}(A) \leq H_2$.

LEMMA 6.2. *Suppose that H_1 and H_2 are maximal subgroups of G that both contain a conjugate of z and suppose that H_1 infects H_2. Let $\sigma := \pi(F(H_1))$. Then the following hold:*

(1) $Z(F(H_1))E(H_1) \leq H_2$.
(2) $[E(H_1), O_q(H_2)] = 1$ for all $q \in \sigma$.
(3) *If* $E(H_1) \neq 1$ *or* $|\sigma| \geq 2$, *then* $F_\sigma(H_2) \leq H_1$.

PROOF. By hypothesis, there exists an involution in $z^G \cap H_1$. This involution is isolated in G because it is conjugate to z, so Corollary 4.8 yields that H_1 is primitive. Similarly H_2 is primitive.

(1) Let $A \leq F(H_1)$ be such that $AC_{F^*(H_1)}(A) \leq H_2$. Then
$$Z(F(H_1))E(H_1) \leq C_{F^*(H_1)}(A) \leq H_2.$$

(2) Let $q \in \sigma$ and $Q := Z(O_q(H_1))$. Then $1 \neq Q \trianglelefteq H_1$ and from (1) we know that $Q \leq H_2$. Moreover $N_G(Q) = H_1$ because H_1 is primitive and therefore $C_{O_q(H_2)}(Q) \leq H_1$ normalises $E(H_1)$. Conversely $E(H_1)$, which lies in H_2 by (1), normalises $C_{O_q(H_2)}(Q)$. Hence
$$[C_{O_q(H_2)}(Q), E(H_1)] \leq O_q(H_2) \cap E(H_1) \leq Z(E(H_1))$$
and the Three Subgroups Lemma yields that $[C_{O_q(H_2)}(Q), E(H_1)] = 1$. Then, since $O^q(E(H_1)) = E(H_1)$, we may apply Thompson's $P \times Q$-Lemma. It gives that $E(H_1)$ centralises $O_q(H_2)$ as stated.

(3) If $E(H_1) \neq 1$, then by (2) we have that $F_\sigma(H_2) \leq N_G(E(H_1)) = H_1$. Now suppose that $|\sigma| \geq 2$ and let $p, q \in \sigma$ be distinct. Again let $Q := Z(O_q(H_1))$ and set $P := Z(O_p(H_1))$. Then $1 \neq P \leq H_2$ by (1) and $C_{O_q(H_2)}(Q) \leq H_1$ and therefore

$$[C_{O_q(H_2)}(Q), P] \leq O_q(H_2) \cap P = 1.$$

Once more we apply Thompson's $P \times Q$-Lemma and obtain that $O_q(H_2) \leq C_G(P) \leq H_1$. Repeating this argument for all primes in σ yields the statement. □

The next theorem is essential for the Bender method. The result is, in fact, due to Bender and is usually stated for maximal subgroups of simple groups. We felt that the fact that G is not necessarily simple makes the quotation of theorems for simple groups slightly inconvenient – so rather than doing that and dealing with case distinctions every time, we decided to rephrase Bender's results for our purpose and to give an explicit proof.

THEOREM 6.3 (Infection Theorem). *Suppose that H_1 and H_2 are maximal subgroups of G that both contain a conjugate of z and suppose that H_1 infects H_2. Set $\sigma := \pi(F(H_1))$.*

(1) $F_{\sigma'}(H_2) \cap H_1 = 1$.
(2) *If q is a prime such that $O_q(H_1) \neq 1$ and $char(H_2) = q$, then $char(H_1) = q$.*
(3) *If $H_2 \hookrightarrow H_1$, then $H_1 = H_2$ or there exists a prime q such that $char(H_1) = q = char(H_2)$.*
(4) *If $E(H_2) \leq H_1$ and $\pi(F(H_2)) \subseteq \sigma$, then $H_1 = H_2$ or there exists a prime q such that $char(H_1) = q = char(H_2)$.*
(5) *If H_1 and H_2 are conjugate and $E(H_2) = 1$, then $H_1 = H_2$ or there exists a prime q such that $char(H_1) = q = char(H_2)$.*

PROOF. Let $A \leq F(H_1)$ be such that $AC_{F^*(H_1)}(A) \leq H_2$ and note that, by Lemma 6.2, we have that $Z(F(H_1))E(H_1) \leq H_2$. We use throughout that H_1 and H_2 are primitive subgroups of G, by Corollary 4.8.

(1) First we note that $F := F_{\sigma'}(H_2) \cap H_1$ acts coprimely on $F(H_1)$ and that $[F, AC_{F(H_1)}(A)] \leq F \cap F(H_1) = 1$. Therefore F centralises a centraliser closed subgroup of $F(H_1)$. Hence from Lemma 2.1 (6) we deduce that $[F, F(H_1)] = 1$. But we also know that F and $E(H_1)$ normalise each other and therefore $[F, E(H_1)] \leq Z(E(H_1))$ whence $[F, E(H_1)] = 1$ by the Three Subgroups Lemma. Thus $F \leq C_{H_1}(F^*(H_1)) = Z(F(H_1))$. This yields that $F = 1$ as stated because F is a σ'-group.

(2) Suppose that $char(H_2) = q$ and that $O_q(H_1) \neq 1$. By Lemma 6.2 (2) we have that $E(H_1)$ centralises $O_q(H_2) = F^*(H_2)$. But then

$$E(H_1) \leq C_{H_2}(F^*(H_2)) = Z(F^*(H_2))$$

and thus $E(H_1) = 1$. Now let $P := O_{q'}(Z(F(H_1)))$, let $Q := Z(O_q(H_1))$ and note that $C_{O_q(H_2)}(Q) \leq H_1$ because H_1 is primitive. As $P \leq H_2$ by

Lemma 6.2 (1), we consider the action of $P \times Q$ on $O_q(H_2) = F^*(H_2)$. Then
$$[C_{O_q(H_2)}(Q), P] \leq O_q(H_2) \cap P = 1.$$
Thompson's $P \times Q$-Lemma yields that $[O_q(H_2), P] = 1$. Therefore $P \leq C_{H_2}(F^*(H_2)) = Z(F^*(H_2))$. But then $P = 1$ because P is a q'-group and thus $F(H_1) = F^*(H_1)$ is a q-group.

(3) From Lemma 6.2 (1) we know that $Z(F(H_1)) \leq H_2$ and $Z(F(H_2)) \leq H_1$. Together with (1) this yields that $\pi(F(H_2)) = \sigma$. Again by Lemma 6.2 (1) we have that $E(H_1)$ and $E(H_2)$ are contained in $H_1 \cap H_2$, thus each component of H_1 or H_2 is a component of $H_1 \cap H_2$. First suppose that $F(H_1) = F(H_2) = 1$. If E is a component of H_1 that is not contained in (and then coincides with) a component of H_2, then E centralises $E(H_2) = F^*(H_2)$ and is therefore contained in $Z(F(H_2)) = 1$. This is impossible and therefore this argument shows that $E(H_1) = E(H_2)$. Then Corollary 4.8 yields that $H_1 = H_2$. Therefore we may suppose that $F(H_1) \neq 1$ or $F(H_2) \neq 1$. As $\pi(F(H_2)) = \sigma$, this implies that $F(H_1) \neq 1 \neq F(H_2)$. If there exists a prime q such that one of $F^*(H_2)$ or $F^*(H_1)$ is a q-group, then $\pi(F(H_2)) = \sigma = \{q\}$ by (2). Thus we suppose that neither $F^*(H_2)$ nor $F^*(H_1)$ is of prime characteristic. Then, as $\pi(F(H_2)) = \sigma$, Lemma 6.2 (3) implies that $F(H_1) \leq H_2$ and also $F(H_2) \leq H_1$ because H_2 infects H_1. So we deduce that $F^*(H_1) \leq H_2$ and $F^*(H_2) \leq H_1$. Let $p \in \sigma$ and set $P_1 := O_p(H_1)$ and $P_2 := O_p(H_2)$. Note that $P_1 F(H_2) = P_1 P_2 \times F_{p'}(H_2)$ is nilpotent. It follows from the previous paragraph that $[P_1, O^p(F^*(H_2))] = 1$ and it follows that
$$[P_1, C_{H_2}(P_2)] \leq C_{H_2}(O^p(F^*(H_2))) \cap C_{H_2}(P_2) \leq C_{H_2}(F^*(H_2)) \leq Z(F(H_2)).$$

In particular $P_1 F(H_2)$ is $C_{H_2}(P_2)$-invariant. But then we see that $P_1 P_2 = O_p(P_1 F(H_2))$ is normalised by $C_{H_2}(P_2)$ and therefore $[P_1, O^p(C_{H_2}(P_2))] = 1$. Hence we have that $O^p(C_{H_2}(P_2)) \leq C_{H_1}(P_1)$ and symmetry yields that $1 \neq O^p(C_{H_1}(P_1)) = O^p(C_{H_2}(P_2))$. This implies that $H_1 = H_2$ because H_1 and H_2 are primitive.

(4) If there exists a prime q such that $F^*(H_2)$ is a q-group, then the hypothesis yields that $O_q(H_1) \neq 1$ and then H_1 is of characteristic q as well, by (2). Now suppose that $F^*(H_1)$ is a q-group. Then the hypothesis implies that $\pi(F(H_2)) \subseteq \sigma = \{q\}$ which means that $F(H_2)$ is a (possibly trivial) q-group. Thus the result follows if $E(H_2) = 1$. Now suppose that $E(H_2) \neq 1$. As $E(H_2) \leq H_1$ with Lemma 6.2 (1), we have that
$$[E(H_2), AC_{F^*(H_1)}(A)] \leq E(H_2) \cap F^*(H_1) = E(H_2) \cap O_q(H_1) \leq Z(E(H_2))$$
because H_1 has characteristic q. The Three Subgroups Lemma yields that $E(H_2)$ centralises $AC_{F^*(H_1)}(A)$. We recall that $AC_{F^*(H_1)}(A)$ is a q-group by our hypothesis that $F^*(H_1)$ is a q-group, and therefore we may apply Thompson's $P \times Q$-Lemma to the action of $E(H_2) \cdot AC_{F^*(H_1)}(A)$ on $F^*(H_1)$. Then we see that $[F^*(H_1), E(H_2)] = 1$. It follows that
$$E(H_2) \leq C_{H_1}(F^*(H_1)) = Z(F(H_1))$$
and therefore $E(H_2) = 1$. Thus $1 \neq O_q(H_2) = F^*(H_2)$. Now the proof is almost finished – it remains to consider the case where $F^*(H_1)$ is not a q-group. Then the hypothesis that $\pi(F(H_2)) \subseteq \sigma$, together with Lemma

6.2 (1) and (3), implies that $F^*(H_2) \le H_1$. Therefore $H_2 \hookrightarrow H_1$ and the result follows from (3).

(5) By hypothesis we have that $E(H_2) = 1$ and $\pi(F(H_2)) = \pi(F(H_1))$. Thus (4) yields the result. □

6.2. t-Minimal Subgroups, Pushing Down and Uniqueness Results

DEFINITION 6.4. Suppose that $t \in G$ is an involution and that W is a $C_G(t)$-invariant $2'$-subgroup W of G. Then W is said to be **t-minimal** if and only if W is minimal with respect to being normalised by $C_G(t)$, but not centralised by t.

LEMMA 6.5. *Suppose that $t \in G$ is an involution, that H is a proper subgroup of G containing $C_G(t)$ and that t is isolated in H. Then precisely one of the following holds:*

(1) $t \in Z(H)$ or
(2) *there exists some odd prime $q \in \pi(F(H))$ such that $O_q(H)$ contains a t-minimal subgroup.*

PROOF. Suppose that $t \notin Z(H)$. Then Lemma 4.2 (6) implies that there exists an odd prime $q \in \pi(F(H))$ such that $[O_q(H), t] \ne 1$. As $[O_q(H), t]$ is $C_H(t)$-invariant and hence $C_G(t)$-invariant, we find a t-minimal subgroup inside $[O_q(H), t]$. □

HYPOTHESIS 6.6.
In addition to Hypothesis 4.1, suppose the following:
- *$t \in C$ is an involution.*
- *Whenever $C_G(t)$ is contained in a proper subgroup H of G, then t is isolated in H.*
- *H_t is a maximal subgroup of G such that $C_G(t) \le H_t$. If possible, we choose H_t such that there exists a prime r such that $O_r(H_t) \ne 1 = C_{O_r(H_t)}(t)$.*
- *We set $\pi_t := \pi(F(H_t))$ and if $C_G(t) \ne H_t$, then we let $q \in \pi_t$ be an odd prime such that $O_q(H_t)$ contains a t-minimal subgroup U_t.*

LEMMA 6.7. *Suppose that Hypothesis 6.6 holds and that $C_G(t) < H_t$. Then $U_t = [U_t, t]$ and U_t centralises $C_{F^*(H_t)}(t)$.*

PROOF. If $[U_t, t] < U_t$, then the t-minimality of U_t forces $[U_t, t, t] = 1$. By Lemma 2.1 (2), this means that $[U_t, t] = 1$, which is a contradiction. The Thompson $P \times Q$-Lemma 2.2, applied to the action of $O_q(C_G(t)) \times \langle t \rangle$ on U_t, together with the minimality of U_t, implies that $O_q(C_G(t))$ centralises U_t. Moreover U_t centralises $F_{q'}(H)$ and $E(H)$ and hence $[C_{F^*(H_t)}(t), U_t] = 1$. □

LEMMA 6.8 (Pushing Down Lemma). *Suppose that Hypothesis 6.6 holds and that $t \in H < G$.*

(1) If $F \leq H$ is a nilpotent $C_H(t)$-invariant subgroup, then $[F,t] \leq F(H)$.
(2) $[O(F(H_t)) \cap H, t] \leq F(H)$.
(3) Suppose that $C_G(t) < H_t$. If $U_t \leq H$, then $U_t \leq O_q(H)$.

PROOF. By hypothesis and Lemma 4.2 (2) we have that $[F,t] \leq [H,t] \leq O(H)$. Hence $[F,t]$ is a nilpotent $C_H(t)$-invariant $2'$-subgroup of H. By Lemmas 2.9 and 2.1 (2) we obtain that $[F,t] = [F,t,t] \leq F(H)$, which is (1).

Parts (2) and (3) both follow from (1) because $O(F(H_t)) \cap H$ and U_t are nilpotent $C_H(t)$-invariant subgroups of H and because $U_t = [U_t, t]$ by Lemma 6.7. □

LEMMA 6.9. *Suppose that Hypothesis 6.6 holds and that $C_G(t) < H_t$. Then H_t is the unique maximal subgroup of G containing $N_G(U_t)$.*

PROOF. First we note that U_t is not normal in G by Lemma 4.7. Hence we may choose $N_G(U_t) \leq H \max G$. Then $C_G(t) \leq H$ and $H_t \looparrowright H$. We set $F := F_{\pi'_t}(H)$ and we see that $F \cap C_G(t) \leq F \cap H_t = 1$, by the Infection Theorem (1). Thus t inverts F.

If $F \neq 1$, then our choice of H_t in Hypothesis 6.6 implies that there exists a prime $r \in \pi_t$ such that $O_r(H_t)$ is inverted by t. By Lemma 4.2 (5), applied to t and H_t, we know that r is odd. Let $X := Z(O_r(H_t))$. Then $N_G(X) = H_t$ because H_t is primitive by Corollary 4.8. We also see that $[X,t] = X$ and that X commutes with U_t whence $X \leq H$. Now t is isolated in H by hypothesis and X is a $C_H(t)$-invariant nilpotent subgroup of H, so the Pushing Down Lemma (1) yields that $X = [X,t] \leq O_r(H)$. It follows that $O_{r'}(H) \leq C_G(X) \leq H_t$. But $r \in \pi_t$, so we deduce that
$$F \leq F_{r'}(H) \leq F \cap H_v = 1,$$
which is a contradiction. We conclude that $F = 1$.

In order to apply the Infection Theorem (4), we still need to show that $E(II) \leq H_t$. But this follows immediately from the fact that t is isolated in H and Lemma 4.2 (5). Hence the Infection Theorem yields that $H = H_t$ or $\mathrm{char}(H) = \mathrm{char}(H_t) = q$. In the second case, we assume that $H_t \neq H$ and apply Theorem 2.27:

H_t and H are primitive by Corollary 4.8. Whenever $C_G(t) \in H_1 < G$, then t is isolated in H_1 by Hypothesis 6.6 and therefore $t \in Z^*(H_1)$ by Lemma 4.2 (1). Also H and H_t contain $C_G(t)$ and they both have characteristic q, so it follows that (H, H_t) is a $\langle t \rangle$-special primitive pair of characteristic q of G. Theorem 2.27 implies that $O_q(H_t) \cap H = 1 = O_q(H) \cap H_t$. But $U_t \leq H$ and therefore $U_t = [U_t, t] \leq O_q(H)$ with the Pushing Down Lemma (3). Hence $1 \neq U_t \leq O_q(H_t) \cap O_q(H)$ and this is a contradiction. Consequently $H_t = H$. □

LEMMA 6.10. *Suppose that Hypothesis 6.6 holds and that $C_G(t) < H_t$. Suppose that $1 \neq X \leq F(H_t)$ is a $U_t \langle t \rangle$-invariant subgroup and that H is a maximal subgroup of G containing $N_G(X)$. Then $H_t = H$ or $\mathrm{char}(H_t) = \mathrm{char}(H) = q$.*

In particular, if $|\pi_t| \geq 2$, then $N_G(X) \leq H_t$.

PROOF. By hypothesis, we have that $X \leq F(H_t)$ and $N_G(X) \leq H$. Therefore $H_t \looparrowright H$. We also have that $U_t \langle t \rangle \leq H$ and thus the Pushing Down Lemma (3) yields that $U_t \leq O_q(H)$. From Lemma 6.9 we know that $N_G(U_t) \leq H_t$, so H infects H_t and the Infection Theorem 6.3 (3) yields the first conclusion. If $|\pi_t| \geq 2$, then only the possibility $H_t = H$ is left and hence $N_G(X) \leq H_t$. □

LEMMA 6.11. *Suppose that Hypothesis 6.6 holds and that $t \neq z$. Suppose that $C_G(t) = H_t$ and that $q \in \pi_t$ is such that $O_q(H_t) \not\leq C$.*

If $Q \in \mathrm{Syl}_q(H_t, \langle z, t \rangle)$ and $T \in \mathrm{Syl}_2(C_G(Q), \langle z \rangle)$, then either

(1) $N_G(Q) \not\leq H_t$ and there exists some $c \in N_G(Q) \cap N_G(T) \cap C$ such that $t^c \in Z(T)$ and $t^c \neq t$

or

(2) $N_G(Q) \leq H_t$ and the involutions t and tz are not conjugate. Moreover, in this case, every z-invariant q-subgroup of $C_G(tz)$ is centralised by z.

PROOF. First we suppose that $N_G(Q) \not\leq H_t$ in order to obtain (1). By hypothesis t centralises Q whilst $C_G(Q) \leq C_G(O_q(H_t)) \leq H_t$, because H_t is primitive by Corollary 4.8. In particular $t \in Z(C_G(Q))$ and therefore $t \in Z(T)$.

Assume that $N_G(Q) \cap N_G(T) \leq H_t$. Then a Frattini argument yields that
$$N_G(Q) = C_G(Q)(N_G(Q) \cap N_G(T)) \leq H_t,$$
which is a contradiction. Thus $N_G(Q) \cap N_G(T) \not\leq H_t$.

By choice of T and Lemma 3.1 (2), we know that $[T, z] = 1$, but $z \notin T$ because z does not centralise Q by hypothesis. We have seen that $N_G(T) \not\leq C_G(t)$, so some element from $N_G(T) \cap N_G(Q)$ maps t to another involution in $Z(T)$. Lemma 3.1 (9) implies that this element can be chosen in C. This proves (1).

For (2) we suppose that $N_G(Q) \leq H_t$ and we set $w := tz$. Let H_w denote a maximal subgroup of G containing $C_G(w)$. As $N_G(Q) \leq H_t$, it follows that $Q \in \mathrm{Syl}_q(G)$ and we recall that Q is centralised by t, but not by z. Lemma 3.10 implies that C does not contain any Sylow q-subgroup of G. Then from Lemma 3.12 we obtain that $|C_K(w)|_q = 1$ and $|C_K(t)|_q \neq 1$. If t and w are conjugate in G, then they are conjugate in C (by Lemma 3.1 (9)) and then the subsets $C_K(t)$ and $C_K(w)$ are C-conjugate as well. This is impossible.

Now let Q_0 be some z-invariant q-subgroup of $C_G(w)$. The previous paragraph implies that $C_C(w)$ already contains a full Sylow q-subgroup Q_1 of $C_G(w)$. With Lemma 4.2 (4) let $Q_0 \leq Q_2 \in \mathrm{Syl}_q(C_G(w), z)$ and with Lemma 3.10, applied to $C_G(w)$, let $x \in C_C(w)$ be such that $Q_1^x = Q_2$. Then $Q_0 \leq Q_2 = Q_1^x \leq C_C(w)$ and hence z centralises Q_0. □

LEMMA 6.12. *Suppose that Hypothesis 6.6 holds, that $t \neq z$ and that C is a maximal subgroup of G. Let $\pi := \pi(F(C))$. If $O(F(C)) \cap H_t \neq 1$, then $[H_t, z]$ is contained in $F_{\pi'}(H_t)$.*

PROOF. Define $X := [H_t, z]$. Then by Lemma 4.11 we have that X is a π'-group and therefore it is sufficient to show that X is nilpotent. We set $D := O(F(C)) \cap H_t$ and see that $D \times \langle z \rangle$ acts coprimely on X. As
$$[C_X(z), D] \leq C_X(z) \cap O(F(C)) = 1,$$
it follows that $C_X(z) \leq C_X(D)$. This means that Theorem 2.24 is applicable: it yields that $[C_X(D), z]$ is normal in X and that $[X, D]$ is a nilpotent normal subgroup of X. With Lemma 4.7 let H be a maximal subgroup of G containing $C_G(D)$. Then $z \in H$ and Lemma 4.2 (1) implies that $H_0 := [H, z] \leq O(H)$. So H_0 is a soluble π'-group, again by Lemma 4.11. Moreover $C \hookrightarrow H$ because $D \leq F(C)$. With the Infection Theorem (1) we deduce that $C \cap H_0 \leq C \cap F_{\pi'}(H) = 1$ and that,

therefore, H_0 is inverted by z. We deduce that H_0 is an abelian normal subgroup of H and in particular $H_0 \leq F(H)$. Now we have that

$$[C_X(D), z] \leq [C_G(D), z] \cap C_X(D) \leq H_0 \cap C_X(D) \leq F(H) \cap C_X(D) \leq F(C_X(D)).$$

Hence $[C_X(D), z]$ is nilpotent, it is normal in X by the previous paragraph, and thus $[C_X(D), z] \leq F(X)$. With Lemma 2.1 (2), it follows that $X = C_X(D)[X, D]$ and finally

$$X = [X, z] \leq [C_X(D), z][X, D, z] \leq F(X),$$

so in particular X is nilpotent. □

CHAPTER 7

Maximal Subgroups Containing C

The objective of this section is the proof of Theorem **A**, and the material is to a large extent taken from [**Wal09**]. We proceed by way of contradiction and begin by phrasing a suitable working hypothesis.

HYPOTHESIS 7.1. *Suppose that Hypothesis 4.1 holds and that $C < M \, max \, G$.*

- *If possible, then choose M such that there exists a prime r such that $O_r(M) \neq 1 = C_{O_r(M)}(z)$.*
- *Set $\pi := \pi(F(M))$ and let $p \in \pi$ be such that $O_p(M)$ contains a z-minimal subgroup U.*
- *Suppose that M is not of characteristic p.*
- *Let $P \in Syl_p(M, z)$ and $Z := \Omega_1(Z(P))$.*

LEMMA 7.2. *Suppose that Hypothesis 7.1 holds. Then $Z \nleq O_p(M)$. In particular Z is not cyclic.*

PROOF. By Hypothesis 7.1 and Lemma 4.12 we have that $P \notin \mathrm{Syl}_p(G, z)$ and therefore $N_G(P)$ is not contained in M. Assume that $Z \leq O_p(M)$. As $[U, Z] = 1$ and Z is a non-trivial z-invariant subgroup of $F(M)$, Lemma 6.10 is applicable to Z. From Hypothesis 7.1 we know that M is not of characteristic p, so the lemma yields that $N_G(Z) \leq M$. Then it follows that $N_G(P) \leq N_G(Z) \leq M$, which is a contradiction. For the second assertion we assume that Z is cyclic. Then $|Z| = p$ and since $Z \cap O_p(M) \neq 1$, this means that $Z \leq O_p(M)$. However this contradicts the first statement. □

LEMMA 7.3. *Suppose that Hypothesis 7.1 holds and that X is a subgroup of $O_p(M)$ such that $1 \neq X = [X, z]$. Then M is the unique maximal subgroup of G containing $N_G(X)$.*

PROOF. By Lemma 4.7 and since X is z-invariant, there exists a maximal subgroup H of G containing $N_G(X)$. As $X \leq O_p(M)$, we have that $M \hookrightarrow H$. Our objective is to apply the Infection Theorem (4) and so we first note that z centralises $E(H)$ by Lemma 4.2 (5). Therefore $E(H) \leq C \leq M$ and next we consider $F := F_{\pi'}(H)$. The Infection Theorem (1) gives that $F \cap M = 1$ and in particular $F \cap C = 1$. Thus F is inverted by z.

(1) $[F, X] = 1$.

PROOF. We note that p is odd which means that we can apply the Pushing Down Lemma 6.8 (2) . It yields that
$$X = [X, z] \leq [O(F(M)) \cap H, z] \leq F(H)$$
and therefore $X \leq O_p(H)$. Then $[F, X] \leq O_{\pi'}(H) \cap O_p(H) = 1$. □

(2) If $r_p(C_C(O_p(M))) \geq 2$, then $F = 1$.

PROOF. Suppose that there exists an elementary abelian p-subgroup $W \leq C_C(O_p(M))$ of order at least p^2. We note that W and z both lie in $N_G(X)$ and hence in H.

Let $w \in W^\#$. Then $z \in C_G(w)$ whence Lemma 4.7 implies that $C_G(w)$ is a proper subgroup of G. Let L be a maximal subgroup of G containing $C_G(w)$. As $W \leq C_C(O_p(M)) \leq C_C(U)$, it follows that $U\langle z \rangle \leq C_G(w) \leq L$. The Pushing Down Lemma (3) implies that $U \leq O_p(L)$. As Hypothesis 7.1 implies Hypothesis 6.6, Lemma 6.9 yields that L infects M. We observed above that the subgroup F is inverted by z, in particular $C_F(w) \leq L$ is inverted by z. It follows that
$$C_F(w) = [C_F(w), z] \leq [L, z] \leq O(L)$$
because $z \in Z^*(L)$ by Lemma 4.2 (1). Since $X \leq C_G(w) \leq L$, we also have that $X = [X, z] \leq O(L)$.

We recall that F centralises X by (1) and that F is a p'-group because $p \in \pi$. Together with Lemma 2.8 we obtain that
$$C_F(w) = C_{C_F(w)}(X) \leq O_{p'}(C_{O(L)}(X)) \leq O_{p'}(O(L)) \leq O_{p'}(L).$$

As $U \leq O_p(L)$, it follows that $[U, C_F(w)] = 1$. By Lemma 2.1 (4) and since W is not cyclic, we have that $F = \langle C_F(w) \mid w \in W^\# \rangle$ and thus $[U, F] = 1$. Now Lemma 6.9 implies that $F \leq N_G(U) \leq M$ and therefore $F = F \cap M = 1$. □

(3) If $[Z, z] \neq 1$, then $F = 1$.

PROOF. Suppose that Z possesses an element $w \neq 1$ that is inverted by z. We noted in the first paragraph of the proof that F is also inverted by z. But we also know that $w \in Z \leq C_G(X) \leq H$ which implies that F is w-invariant. We conclude that F is centralised by $\langle w \rangle = [\langle w \rangle, z]$. Now let L be a maximal subgroup of G containing $N_G(\langle w \rangle)$. Then z, X, U, Z and – as we have just seen – also F are contained in L. The Pushing Down Lemma (2) and (3) imply that X and U are both contained in $O_p(L)$ and hence in $O_p(C_G(w))$. From Lemma 4.2 (1) we know that $z \in Z^*(L)$ and therefore F, which is inverted by z, lies in $O(L)$. Lemma 2.8 gives that
$$F \leq O_{p'}(C_G(X)) \cap C_{O(L)}(w) \leq O_{p'}(C_{O(C_G(w))}(X)) \leq O_{p'}(O(C_G(w))).$$

As $U \leq O_p(C_G(w))$, it follows that $[U, F] = 1$ and therefore $F \leq C_G(U) \leq M$, with Lemma 6.9. Then $F = F \cap M = 1$ as stated. □

Now we can finish the proof. We know that Z is elementary abelian of order at least p^2, by Lemma 7.2. If $[Z, z] = 1$, then $Z \leq C_C(O_p(M))$ whence (2) is applicable and gives that $F = 1$. If $[Z, z] \neq 1$, then (3) implies that $F = 1$.

It follows that $F(H)$ is a π-group. The Infection Theorem (4) gives that $H = M$ or that H and M both have characteristic p. As M is not of characteristic p by Hypothesis 7.1, we conclude that $H = M$. Therefore $N_G(X) \leq M$. □

The next lemmas include a special case of the situation where a component of M is isomorphic to $PSL_2(q)$ for some odd number q, that is needed in Section 14.

LEMMA 7.4. *Suppose that Hypothesis 7.1 holds and that $E(M) = 1$ or that there exists an odd number q such that $E(M) \simeq PSL_2(q)$. Suppose further that W is an elementary abelian subgroup of M of order p^2 that is centralised or inverted by z. Then z inverts W and $[C_{O_p(M)}(W), z] = 1$. In particular $C_P(z)$ is cyclic.*

PROOF. We first note that Hypothesis 7.1 implies Hypothesis 6.6 and therefore the results from Chapter 6 are applicable.

Assume that the assertion in the lemma is wrong and choose an elementary abelian subgroup W of M of order p^2 such that z centralises W or $[C_{O_p(M)}(W), z] \neq 1$. We recall that, by Hypothesis 7.1, we know that $O_p(M)$ is not centralised by z. Hence if $[W, z] = 1$, then Thompson's $P \times Q$-Lemma, applied to the action of $W \times \langle z \rangle$ on $O_p(M)$, yields that $[C_{O_p(M)}(W), z] \neq 1$. For the remainder of the proof we therefore assume that $[C_{O_p(M)}(W), z] \neq 1$ and we work towards a contradiction.

As W is z-invariant, Lemma 4.2 (4) implies that W is contained in a z-invariant Sylow p-subgroup of M, so we may suppose that $W \leq P$. From Lemma 4.12 we know that $P \notin \mathrm{Syl}_p(G)$ and in particular $N_G(P) \not\leq M$. Lemma 3.7 yields that

$$N_G(P) = C_{N_G(P)}(z)(K \cap N_G(P))$$

and this implies that $K \cap N_G(P) \not\subseteq M$. Let $h \in K \cap N_G(P)$ be such that $h \notin M$ and let $g \in G$ be such that $h = zz^g$. Then $M^h = M^{z^g}$ and $t := z^g$ is an involution in $z^G \cap N_G(P)$ that is not contained in M and hence does not normalise M (because M is primitive by Corollary 4.8). If M has a component, then we denote it by E and we recall that $E(M) = E$ by hypothesis. Our assumption that $[C_{O_p(M)}(W), z] \neq 1$ implies that $C_{O_p(M)}(W)$ possesses an element x of order p that is inverted by z.

(1) Suppose that $y \in O_p(M)$ is a non-trivial element that is inverted or centralised by z. Then $N_G(\langle y \rangle) \leq M$. In particular $C_G(x) \leq M$.

PROOF. If y is centralised by z, then $y \in C_{O_p(M)}(z) \leq C_G(U)$ by Lemma 6.7 and thus $N_G(\langle y \rangle) \leq M$ by Lemma 6.10. If y is inverted by z, then

$$1 \neq \langle y \rangle = [\langle y \rangle, z] \leq O_p(M)$$

and hence $N_G(\langle y \rangle) \leq M$ by Lemma 7.3. As z inverts x, it follows that $C_G(x) \leq N_G(\langle x \rangle) \leq M$. □

(2) $F(M^t) \leq M$ and, for all $w \in W^\#$, we have that $x \in O_p(C_G(w))$.

PROOF. First we see that $O_p(M^t) \leq P^t = P \leq M$. Then we consider $Q := O_{p'}(M^t)$ and the coprime action of W on it. Lemma 2.1 (4) implies that

$$Q = \langle C_Q(w) \mid w \in W^\# \rangle.$$

With the Pushing Down Lemma (2) we see, for all $w \in W^\#$, that

$$\langle x \rangle = [\langle x \rangle, z] \leq [O_p(M) \cap N_G(\langle w \rangle), z] \leq O_p(N_G(\langle w \rangle)).$$

In particular $x \in O_p(C_G(w))$ as stated. Also $x \in P \leq M^t$, so as x acts on Q, it follows that
$$[x, C_Q(w)] \leq O_p(C_G(w)) \cap Q = 1.$$
We deduce that $C_Q(w) \leq C_G(x) \leq M$ by (1) and thus $Q \leq M$. □

(3) $E(M^t) \not\leq M$, so in particular $E(M) \neq 1$.

PROOF. Otherwise, together with (2), we see that $F^*(M^t) \leq M$ and hence $M^t \looparrowright M$. As t is an involution, we also have that $M \looparrowright M^t$ and therefore the Infection Theorem (3) yields that $M = M^t$ or that M and M^t have the same prime characteristic. We know that $F^*(M) \neq O_p(M)$ by Hypothesis 7.1, so we deduce that $M = M^t$ and this is a contradiction. □

(4) Let s be an involution in E. Then $|C_E(s)|$ is divisible by p.

PROOF. Set $D := O_p(C_G(s)) \cap O_p(M)$. First we note that $D \neq 1$ because $U \leq C_G(E) \leq C_G(s)$ and then the Pushing Down Lemma (3) yields that $U \leq O_p(C_G(s))$ and hence $U \leq D$. As $D \leq O_p(M)$ and M is not of characteristic p, Lemma 6.10 forces $N_G(D) \leq M$. If $D = O_p(C_G(s))$, then D is $C_G(s)$-invariant and therefore $C_G(s) \leq M$. But $s \neq z$ because $z \notin E$ and thus Lemma 4.13 forces $C_G(s) \not\leq M$, which is a contradiction.

Therefore $D \neq O_p(C_G(s))$ and $D < Y := N_{O_p(C_G(s))}(D)$. Then $Y \leq N_G(D) \leq M$ and hence Y acts on E and leaves the dihedral group $C_E(s)$ invariant. (Recall that $E \simeq PSL_2(q)$.) Conversely $C_E(s) \leq C_G(s) \cap M \leq N_G(Y)$.

Now we assume that $C_E(s)$ is a p'-group. Then $[Y, C_E(s)] \leq Y \cap C_E(s) = 1$ and it follows that $Y/C_Y(E)$ cannot induce a non-trivial field automorphism on E. Next we argue that $D = C_Y(E)$: It is immediate that
$$D \leq Y \cap O_p(M) \leq Y \cap C_G(E)$$
and for the other inclusion we note that $C_G(E) \leq M$, because M is primitive, and hence
$$C_Y(E) \leq O_p(C_G(s)) \cap C_G(E) \cap N_G(D) \leq O_p(C_G(s)) \cap C_M(E)$$
$$\leq O_p(M) \cap O_p(C_G(s)) = D.$$
Consequently Y/D induces a non-trivial inner automorphism on E centralising s and it follows that p divides $|C_E(s)|$. This contradicts our assumption that $C_E(s)$ is a p'-group. Hence $|C_E(s)|$ is divisible by p as stated. □

Next we argue that the action of W on E^t is faithful:

Assume otherwise and let $y \in W^\#$ be such that $[E^t, y] = 1$. Then $x \in O_p(C_G(y))$ by (2) and therefore
$$[E^t, x] \leq E^t \cap O_p(C_G(y)) \leq O_p(E^t) = 1,$$
contrary to (1) and (3).

Thus W acts faithfully on E^t and it follows from (4) that the subgroup W_0 of W inducing inner automorphisms on E^t has order p. Each element from $W \backslash W_0$ induces a field automorphism on E^t of order p. Hence there exists some $w \in W \backslash W_0$ such that $x \notin \langle w \rangle C_{M^t}(E^t)$. Then $O_p(C_{E^t}(w)) = 1$ and it follows with (2) that

$$[C_{E^t}(w), x] \leq C_{E^t}(w) \cap O_p(C_G(w)) \leq O_p(C_{E^t}(w)) = 1.$$

We deduce that $E^t = \langle C_{E^t}(w), C_{E^t}(x) \rangle \leq C_G(x)$, which is a contradiction to (1) and (3). This completes the proof of the lemma. □

LEMMA 7.5. *Suppose that Hypothesis 7.1 holds and that $E(M) = 1$ or that there exists an odd number q such that $E(M) \simeq PSL_2(q)$. Then*
  (1) *$C_Z(z)$ has order p and it is the unique subgroup of P of order p that is centralised by z;*
  (2) *$I_Z(z)$ has order p and it is the unique subgroup of P of order p that is inverted by z;*
  (3) *$|Z| = p^2$ and $Z = \Omega_1(P)$;*
  (4) *z inverts $O_p(M)$ and*
  (5) *$O_p(M)$ is cyclic.*

PROOF. We recall that $O_p(M) \cap Z \neq 1$ and that $|Z| \geq p^2$ by Lemma 7.2. Moreover Lemma 2.1 (2) yields that

$$Z = C_Z(z) \times [Z, z] = C_Z(z) \times I_Z(z).$$

(1) Lemma 7.4 gives that $C_Z(z)$ has at most order p and that (therefore) Z is not centralised by z. If $C_Z(z) = 1$, then z inverts Z whence Lemma 7.4 implies that $[C_{O_p(M)}(Z), z] = 1$. But $Z \leq Z(P)$ is centralised by $O_p(M)$. So $[O_p(M), z] = 1$, which is a contradiction.

We conclude that $C_Z(z)$ has order p and is, by Lemma 7.4, the unique subgroup of $C_P(z)$ of order p.

(2) As $|Z| \geq p^2$, it follows from (1) that $I_Z(z) \neq 1$. Assume that $I_Z(z)$ possesses a subgroup V of order p^2. Then, as $V \leq Z(P) \leq C_M(O_p(M))$, it follows with Lemma 7.4 that $[O_p(M), z] = [C_{O_p(M)}(V), z] = 1$, which is a contradiction. It is left to show that $I_Z(z)$ is the only subgroup of P of order p that is inverted by z.

First assume that $Y \leq O_p(M)$ is distinct from $I_Z(z)$, has order p and is inverted by z. Then $W := Y I_Z(z)$ is elementary abelian of order p^2 and we may apply Lemma 7.4. Thus $[Y, z] \leq [C_{O_p(M)}(W), z] = 1$, which is a contradiction. We emphasise here that this means that $I_Z(z) \leq O_p(M)$. Next assume that $Y_1 \leq P$ is distinct from $I_Z(z)$, has order p and is inverted by z. Then $W_1 := Y_1 I_Z(z)$ is elementary abelian of order p^2, and Lemma 7.4 yields that $[I_Z(z), z] \leq [C_{O_p(M)}(W_1), z] = 1$. This is impossible again.

(3) Statements (1) and (2) imply that $|Z| = p^2$, and we know that $Z \leq \Omega_1(P)$. Assume that $Z \neq \Omega_1(P)$. Then there exists a subgroup of P of order p that is not contained in Z. As $Z \leq Z(P)$, it follows that $r(P) \geq 3$. But P is z-invariant and therefore, with Lemma 2.6, there exists a z-invariant elementary abelian subgroup X of P of order p^3. By Lemma 2.1 (2), we have that $X = C_X(z) \times [X, z]$. But then either $C_X(z)$ or $I_X(z)$ has order at least p^2 and that contradicts (1) and (2) above. Thus $\Omega_1(P) \leq Z$.

(4) From Hypothesis 7.1 we know that $O_p(M)$ is not centralised by z and hence there exists a subgroup J of $O_p(M)$ of order p that is inverted by z. Then (2) implies that $J = I_Z(z)$. As $Z \not\leq O_p(M)$ with Lemma 7.2, it follows that $C_Z(z) \not\leq O_p(M)$. But then (1) yields that $O_p(M)$ does not contain any subgroup of order p that is centralised by z. Hence $C_{O_p(M)}(z) = 1$.

(5) We know from (4) that z inverts $O_p(M)$ whence $O_p(M)$ is abelian. Then (2) yields that $O_p(M)$ contains a unique subgroup of order p and, as p is odd, this means that $O_p(M)$ is cyclic. □

LEMMA 7.6. *Suppose that Hypothesis 7.1 holds and that $E(M) = 1$ or that there exists an odd number q such that $E(M) \simeq PSL_2(q)$. Then*

$$U \leq \bigcap_{g \in G} M^g.$$

PROOF. Let $g \in G \backslash M$, let $t \in z^G \cap M^g$ and set $D := M \cap M^t$. We note that, by Lemma 3.1 (8) applied to t, the involution t is not contained in M. From Hypothesis 7.1 we know that z does not centralise $O_p(M)$ and hence $[P, z] \neq 1$. Then Lemmas 3.13 and 4.9 (4) yield that p divides $|M : C| = |D : C_D(t)|$ and it follows that $|[D, t]| / |[D, t] \cap C_D(t)|$ is divisible by p. Thus there exists a subgroup X of D of order p that is inverted by t. We show that X is conjugate to U.

By Lemma 7.5 (3) we have that $Z = \Omega_1(P)$. Lemma 4.7 implies that $N_G(Z)$ is a proper subgroup of G. Then we let $Q \in \mathrm{Syl}_p(N_G(Z), z)$ (with Lemma 4.2 (4)) be such that $P \leq Q$. From Lemma 7.5 (5) we know that $|U| = p$ and $U \triangleleft M$. Thus $N_G(U) = M$ by Corollary 4.8 and it follows that $P = N_Q(U)$. Let $Q_0 := N_Q(P)$. Then $Q_0 \not\leq M$ because $P \notin \mathrm{Syl}_p(G)$ by Lemma 4.12. In particular Q is not contained in M. As $|Q_0 : N_{Q_0}(U)| = |Q_0 : P| \geq p$, we see that U^{Q_0} has at least p elements. But Q_0 normalises Z and therefore every element of U^{Q_0} is one of the p+1 subgroups of Z of order p. We deduce that U^{Q_0} has precisely p elements and that Q_0 normalises (and therefore centralises) a subgroup Y of order p of Z. The subgroup $C_Z(Q_0)$ is z-invariant and hence coincides with U or with $C_Z(z)$, because these two subgroups are the unique z-invariant subgroups of Z of order p. But $C_Z(Q_0) \neq U$ because $N_G(U) = M$ and $Q_0 \not\leq M$. It follows that $C_Z(Q_0) = C_Z(z)$.

Now we recall that our objective is to show that X is conjugate to U: We know that $X \leq D \leq M$. Therefore, by Lemma 7.5 (3) and Sylow's Theorem, the subgroup X is conjugate in M to a subgroup of order p in Z, i.e. to a member of U^{Q_0} or to $C_Z(z)$.

First suppose that X is not conjugate to U. Then X is conjugate to $C_Z(z)$. Thus there exists a conjugate t' of z distinct from t such that t' and t are both contained in $N_G(X)$. Hence they are conjugate in $N_G(X)$ by Lemma 3.1 (4), applied to the isolated involution t. This is impossible because t' centralises X whereas t inverts it. Thus X must be conjugate to U. Now let $h \in G$ be such that $X = U^h$. Then $t \in N_G(X) = N_G(U^h) = M^h$. As $t \in M^g$ and as every conjugate of z is contained in a unique conjugate of M, by Lemma 3.1 (8), this yields that $M^h = M^g$. Now we see that $X \leq D \leq M$ normalises $U \triangleleft M$ and therefore $[X, U] = 1$. So we have that $U \leq N_G(X) = M^g$. As $g \in G \backslash M$ was arbitrary, it follows that U lies in every conjugate of M in G, as stated. □

PROOF OF THEOREM **A**.

Recall that in Theorem **A**, we suppose that G is a minimal counter-example to the Z*-Theorem, so Hypothesis 4.1 is satisfied. Assume that Theorem **A** fails. Then C is properly contained in M and Lemma 6.5 implies that there exists an odd prime p such that $O_p(M)$ contains a z-minimal subgroup U. Again by the failure of Theorem **A**, we have that $F^*(M) \neq O_p(M)$ and $E(M) = 1$. In particular Hypothesis 7.1 is satisfied and we may apply Lemmas 7.2 up to 7.5. The (by Lemma 7.5 (2) unique) subgroup of P of order p that is inverted by z must now be U, by Lemma 7.5 (5). Lemma 7.6 implies that U is contained in every conjugate of M in G. Then
$$1 \neq U \leq \bigcap_{g \in G} M^g \trianglelefteq G,$$
but this contradicts Lemma 4.7. Therefore Theorem **A** is proved. □

The Bender method returns to the scene in the proof of the next few results. They lead to a new proof of Theorem 6.19 in [**Wal09**], avoiding difficulties when quoting results from [**BG94**]. We would like to point out that some of the arguments in this new proof follow ideas from Sections 7 and 8 in Chapter II of [**BG94**]. Here we only suppose that our main hypothesis (4.1) holds.

LEMMA 7.7. *Suppose that C is a maximal subgroup of G and that $Y_0 \leq F(C)$ is centraliser closed in $F(C)$. Let $Y := Y_0 E(C)$ and $\pi := \pi(F(C))$. Then the following hold:*

(1) *For all $r \in \pi$, we have that $C_G(O_r(Y_0)) \leq C$. In particular $C_G(Y_0)$ and (therefore) $C_G(Y)$ are contained in C.*
(2) *$C_G(Y)$ is a π-group.*
(3) *$\mathcal{U}_C(Y, \pi') = \{1\}$.*

Suppose that $Y \leq H \max G$. Then the following hold:

(4) *$H = C_H(z) F_{\pi'}(H)$.*
(5) *For all $\sigma \subseteq \pi'$, we have that $\langle \mathcal{U}_H(Y, \sigma) \rangle = F_\sigma(H)$.*
(6) *For all $q \in \pi'$, the subgroup $O_q(H)$ is the unique maximal Y-invariant q-subgroup of H.*

PROOF.

(1) As Y_0 is centraliser closed in $F(C)$, we have that $Z(F(C)) \leq Y_0$. Hence for all primes $r \in \pi$, Corollary 4.8 implies that
$$C_G(O_r(Y_0)) \leq C_G(Z(O_r(C))) \leq N_G(Z(O_r(C))) = C.$$
Then it follows of course that $C_G(Y) \leq C_G(Y_0) \leq C_G(O_r(Y_0)) \leq C$.

(2) Suppose that $x \in C_G(Y)$ is a π'-element. Then $x \in C$ by (1) and therefore x acts coprimely on $F(C)$ and centralises a centraliser closed subgroup, namely Y_0. Lemma 2.1 (6) implies that x centralises $F(C)$. Moreover $x \in C_G(Y) \leq C_G(E(C))$ which forces $x \in C_C(F^*(C)) \leq Z(F(C))$. But this is a π-group, so $x = 1$.

(3) Let $X \in \mathcal{U}_C(Y, \pi')$. Then $[X, Y_0] \leq X \cap F(C) = 1$, so X centralises the centraliser closed subgroup Y_0 of $F(C)$. With Lemma 2.1 (6) we deduce that $[X, F(C)] = 1$. Moreover $[X, E(C)] \leq X \cap E(C) \trianglelefteq E(C)$. The hypothesis $C = M$ implies that $2 \in \pi$ and therefore the set π' consists

of odd primes. As components are not soluble, they have even order. In particular $X \cap E(C)$ cannot be a product of components of C. We conclude that $X \cap E(C) \leq Z(E(C))$ and then that X centralises $E(C)$, by the Three Subgroups Lemma. Now we have that $X \leq C_C(F^*(C)) = Z(F(C))$ whence $X = 1$.

Statements (4)–(6) are immediate from (3) if $H = C$, so we suppose from now on that $H \neq C$.

(4) The hypothesis yields that $C_{F^*(C)}(Y_0) \leq Y \leq H$ and thus $C \looparrowright H$. Moreover $z \notin Z(H)$ because $H \neq C$. With Lemma 4.2 (5) we have that $E(H) \leq C$ and also Lemma 4.10 implies that all z-invariant π-subgroups of H are centralised by z. As $z \notin Z(H)$, this forces, with Lemma 4.2 (6), that $F := F_{\pi'}(H) \neq 1$. With the Infection Theorem (1) we see that $F \cap C = 1$ whence F is inverted by z. It follows that z centralises $E(H)$ and $F_\pi(H)$ and inverts F, so we conclude that
$$[H, z] \leq C_H(F^*(H)) \leq Z(F(H)).$$
From Lemma 2.5 we know that $[H, z]$ is a π'-group, so $[H, z] \leq F$.

(5) Let $X \in \mathsf{M}_H(Y, \sigma)$. As $C_X(z) \in \mathsf{M}_C(Y, \sigma)$ and $\sigma \subseteq \pi'$, part (3) forces $C_X(z) = 1$ and thus z inverts X. By (4) this implies that $X = [X, z] \leq [H, z] \leq F(H)$ and therefore $X \leq F_\sigma(H)$. Conversely $F_\sigma(H)$ is a member of the set $\mathsf{M}_H(Y, \sigma)$.

(6) Let $q \in \pi'$ and let $Q \in \mathsf{M}_H(Y, q)$. Then (5), applied to $\sigma := \{q\}$, yields that $Q \leq O_q(H)$. This means that $O_q(H)$ is the unique maximal member of $\mathsf{M}_H(Y, q)$. □

LEMMA 7.8. *Suppose the following:*
- *C is maximal in G and $Y_0 \leq F(C)$ is centraliser closed in $F(C)$;*
- *$Y := Y_0 E(C)$ and $Y \leq H \max G$, $H \neq C$; and*
- *$q \in \pi(F(C))'$ is such that $O_q(H) \neq 1$.*

Then $O_q(H)$ is the unique maximal Y-invariant q-subgroup of G that intersects H non-trivially.

PROOF. We know from Lemma 7.7 (6) that $\mathsf{M}^*_H(Y, q) = \{O_q(H)\}$. Let $Q_0 := O_q(H) \leq Q \in \mathsf{M}^*_G(Y, q)$ and note that, by Corollary 4.8, we have that $N_G(Q_0) = H$. Therefore $N_Q(Q_0) \leq H$, implying $N_Q(Q_0) = Q_0$ because $N_Q(Q_0)$ is a Y-invariant q-subgroup of H. As Q is a q-group, this forces $Q_0 = Q$. So we have that Q_0 is already a maximal Y-invariant q-subgroup of G.

Assume that our assertion is wrong. Let $\pi := \pi(F(C))$ and choose $Q_1 \in \mathsf{M}^*_G(Y, q)$ such that $Q_1 \cap H \neq 1$, but $Q_1 \neq Q_0$ and such that the intersection $D := Q_1 \cap Q_0$ is as large as possible. Then $Q_1 \cap H$ is a Y-invariant q-subgroup of H and hence lies in Q_0; in fact $Q_1 \cap H = D$ and in particular $D \neq 1$. Let $N_G(D) \leq L \max G$ (with Lemma 4.7) and note that $Y \leq L$. As D is a Y-invariant q-subgroup of L, Lemma 7.7 (6) yields that $D \leq O_q(L)$. We also know that Q_0 and $O_q(L)$ are abelian because these subgroups are inverted by z (by Lemma 7.7 (3)). We deduce first that $Q_0 \leq C_G(D) \leq L$, then, with part (6) of the same lemma, that $Q_0 \leq O_q(L)$, and then that $O_q(L) \leq C_G(Q_0) \leq H$. It is Lemma 7.7 (6) once more that yields that $O_q(L) \leq O_q(H) = Q_0$. Therefore $O_q(L) = O_q(H)$ and $L = H$ by

7. MAXIMAL SUBGROUPS CONTAINING C

Corollary 4.8. We conclude that $N_G(D) \leq H$ and hence $N_{Q_1}(D) \leq Q_1 \cap H = D$, forcing $D = Q_1$. It follows that $Q_1 = Q_0$ after all, which is a contradiction. Thus $O_q(H)$ is the unique maximal Y-invariant q-subgroup of G that intersects H non-trivially. □

THEOREM 7.9. *Suppose that C is maximal in G and that $p \in \pi(F(C))$ is an odd prime such that $O_p(C)$ contains an elementary abelian subgroup X_0 of order p^3. Then for all $x \in X_0^\#$, the unique maximal subgroup of G containing $C_G(x)$ is C.*

PROOF. Let $Y_0 := C_{F(C)}(X_0)$ and $Y := Y_0 E(C)$. Then $X_0 \leq Y_0$, moreover Y_0 is centraliser closed in $F(C)$ and hence the previous lemmas are applicable. We let $\pi := \pi(F(C))$ and note that $|\pi| \geq 2$ because $2 \in \pi$ and p is odd. From Lemma 4.7 we know that $Y \neq G$ and therefore we may choose a maximal subgroup H of G containing Y. Then $C \looparrowright H$ and $z \in H$. In particular Lemma 4.2 (5) implies that $E(H) \leq C$. If $F(H)$ is a π-group, then the Infection Theorem (4) yields that $H = C$ and there is nothing left to prove. Thus we suppose that $H \neq C$ and (hence) that there exists a prime $q \in \pi'$ such that $Q := O_q(H) \neq 1$. By Lemmas 7.7 (3) and 7.8, it follows that z inverts Q and that Q is the unique member of $\mathcal{W}_G^*(Y,q)$ intersecting H non-trivially. We go further now and show that $\mathcal{W}_G^*(Y,q) = \{Q\}$:

Let $Q_1 \in \mathcal{W}_G^*(Y,q)$ be arbitrary. As Q_1 is Y-invariant, the coprime action of X_0 on Q_1 and Lemma 2.1 (4) give that

$$Q_1 = \langle C_{Q_1}(X_1) \mid X_1 \leq X_0, |X_0 : X_1| = p \rangle.$$

Choose $X_1 \leq X_0$ of index p and such that $C_{Q_1}(X_1) \neq 1$. Then X_1 is elementary abelian of order p^2 and so we can use the same argument for the coprime action of X_1 on Q, namely

$$Q = \langle C_Q(x) \mid x \in X_1^\# \rangle.$$

Let $x \in X_1^\#$ be such that $C_Q(x) \neq 1$. Then $z \in C_G(x)$ and therefore Lemma 4.7 implies that $C_G(x) < G$. We let $C_G(x) \leq H_1 \max G$ and we observe that $Y \leq C_G(x) \leq H_1$. Therefore $C_{Q_1}(X_1)$ and $C_Q(x)$ are Y-invariant q-subgroups of H_1 and thus they are contained in $O_q(H_1)$, by Lemma 7.7 (6). In particular $O_q(H_1) \neq 1$ and Lemma 7.8 yields that $O_q(H_1)$ is the unique member of $\mathcal{W}_G^*(Y,q)$ intersecting H_1 non-trivially. But Q and Q_1 intersect H_1 non-trivially, so we deduce that $Q_1 = Q$, i.e. $\mathcal{W}_G^*(Y,q) = \{Q\}$. It follows that $O_q(H) = Q = O_q(H_1)$ and thus $H_1 = H$, because H and H_1 are primitive by Corollary 4.8.

Let $F := N_{F^*(C)}(Y)$. Then F leaves Q invariant which means that $Q \in \mathcal{W}_G(F,q)$. As $\mathcal{W}_G(F,q) \subseteq \mathcal{W}_G(Y,q)$, it follows that, conversely, every member of $\mathcal{W}_G(F,q)$ lies in Q. Hence $\mathcal{W}_G^*(F,q) = \{Q\}$. But Y is subnormal in $F^*(C)$, so this argument shows that $\mathcal{W}_G(F^*(C),q) = \{Q\}$. However, this implies that $Q = O_q(H)$ is C-invariant and thus $C = H$ contradicting our assumption.

We established that $Y = C_{F^*(C)}(X_0)$ lies in a unique maximal subgroup of G, namely in C. If $x \in X_0^\#$, then $C_{F^*(C)}(X_0) \leq C_G(x) < G$ and therefore $C_G(x)$ lies in the unique maximal subgroup of G containing $C_{F^*(C)}(X_0)$, i.e. in C. □

CHAPTER 8

The 2-rank of $O_{2',2}(C)$

In this section we prove Theorem **B** by analysing the behaviour of involutions from $O_{2',2}(C)$. Some of the arguments from the first part of this chapter appear again towards the endgame. For example in Theorem 8.10, one of the main results of this section, we basically present the proof given in [**Wal09**], but the reader will notice that similar, more complicated arguments are used when we begin to analyse maximal subgroups containing the centraliser of an involution from $C\backslash O_{2',2}(C)$.

8.1. Involutions in $O_{2',2}(C)\backslash\{z\}$

We begin with a special hypothesis and corresponding notation. The objective of the first part of this chapter is to exactly understand what centralisers of involutions distinct from z in $O_{2',2}(C)$ look like – if such involutions exist.

HYPOTHESIS 8.1. *In addition to Hypothesis 4.1, suppose that $a \in O_{2',2}(C)$ is an involution distinct from z. Moreover*

- $V := \langle a, z \rangle$ and $b := az$. Let $V \leq S \in Syl_2(G)$.
- For all $v \in \{a, b\}$ let $C_G(v) \leq H_v \, max \, G$ with $\pi_v := \pi(F(H_v))$ and such that, if possible, there exists a prime $r_v \in \pi_v$ with $C_{O_{r_v}(H_v)}(v) = 1$.
- Let $C \leq M \, max \, G$ and $\pi := \pi(F(M))$ and, if possible, choose M such that there exists a prime $r \in \pi$ with $C_{O_r(M)}(z) = 1$.
- For all $v \in \{a, b\}$, if $C_G(v) \neq H_v$, then let U_v denote a v-minimal subgroup of G that is contained in $F(H_v)$ and if $C \neq M$, then let U be a z-minimal subgroup of G contained in $F(M)$.
- If a and b are conjugate, then suppose that H_a and H_b are conjugate.

LEMMA 8.2. *Suppose that Hypothesis 8.1 holds. Then*
(1) $r_2(C_G(V)) = 2$;
(2) *if $a, z \in H \leq G$, then a and b are either conjugate or isolated in H; and*
(3) *a and b are isolated in every proper subgroup of G containing $C_K(a)$.*

PROOF. As a and z commute, the subgroup V is elementary abelian of order 4. Then Lemma 5.14 is applicable and it yields that $r_2(C_G(V)) = 2$.

For (2) we suppose that $a, z \in H \leq G$ and that a and b are not conjugate in H. Let $P \in Syl_2(H)$ be such that $V \leq P$. Then Lemma 3.1 (2) implies that $z \in Z(P)$. Hence if $N_P(V)$ does not centralise a, then it interchanges a and b. But a and b are not conjugate in H whence $N_P(V) = C_P(V) = C_P(a)$. Part (1) yields that V is the unique elementary abelian subgroup of order 4 in $C_G(V)$ and therefore V is normalised by $N_P(N_P(V))$. Hence $N_P(N_P(V)) = N_P(V)$ whence $P = N_P(V) = C_P(V)$. In particular $r(P) = 2$ by (1) and it follows that a, z and az

are the only involutions in P. We note that distinct involutions from P are never conjugate in H, so (2) holds.

Now we turn to (3) and we suppose that $a, z \in H < G$ and that $C_K(a) \subseteq H$. If a and b are not isolated in H, then (2) says that they are conjugate in H. So by Lemma 3.1 (9) they are conjugate in $C_H(z)$. Let $x \in C_H(z)$ be such that $b = a^x$. Then $C_K(b) = C_{K^x}(a^x) = (C_K(a))^x$ because K is C-invariant by Lemma 3.3 (1). Therefore $C_K(b) \subseteq H$. Theorem 3.6 forces K to be contained in H. As $z \in H$, Lemma 4.4 gives that $H = G$, which is a contradiction. $\square$

LEMMA 8.3. *Suppose that Hypothesis 8.1 holds. Then*
(1) G *is non-abelian and simple;*
(2) $G = \langle C_K(a), C_K(b) \rangle$;
(3) $O^2(C) = C$ *and*
(4) $r_2(G) = 2$.

PROOF. We know from Lemma 4.4 that $G = F^*(G)\langle z \rangle$ and that $F^*(G)$ is non-abelian and simple. In particular $|G : F^*(G)| \leq 2$ and therefore $V \cap F^*(G) \neq 1$. If $G = F^*(G)$, then G is simple as stated. Therefore we assume that $F^*(G) < G$ and in particular that $z \notin F^*(G)$. Lemma 4.2 (1) yields that $F^*(G)$ does not contain any isolated involution, but we know that $F^*(G)$ contains an involution from V and therefore a and b are not isolated in G. With Lemma 8.2 (2) it follows that a and b are conjugate in G. But then a and b are both contained in $F^*(G)$ and hence $z = ab \in F^*(G)$, which is a contradiction.

We turn to (2) and apply Theorem 3.6 together with (1) and the fact that $\langle K \rangle$ is a normal subgroup of G by Lemma 4.3 (3). This yields that $\langle C_K(a), C_K(b) \rangle = \langle K \rangle = G$.

For (3) we note that $O^2(G) = G$ by (1), so $O^2(C) = C$ by Lemma 3.1 (10).

It is left to prove (4). Let $S_0 := S \cap O_{2',2}(C)$. Then $V \leq S_0 \trianglelefteq S$ and in particular $r(S_0) \geq 2$. First we assume that S has no elementary abelian normal subgroup of order 4 that is contained in S_0. In that case Lemma 2.11 implies that S_0 is a dihedral or semi-dihedral group of order at least 8. So we know that $\mathrm{Aut}(S_0)$ is a 2-group, by Lemma 2.14. Recall that $\overline{C} := C/O(C)$. In this factor group we have that $\overline{S_0} = O_2(\overline{C})$ and therefore $[O^2(\overline{C}), \overline{S_0}] = 1$. But $O^2(\overline{C}) = \overline{C}$ by (3) and hence $[\overline{C}, \overline{S_0}] = 1$. Then it follows that $\overline{S_0}$ is abelian and thus S_0 is abelian. But this is not the case and consequently S_0 contains a normal subgroup B of S that is elementary abelian of order 4. With Lemma 8.2 (1), applied to any involution $t \in B$ distinct from z, we deduce that $z \in B$ (because z centralises B) and hence that $r_2(C_G(B)) = 2$.

Suppose that A is an elementary abelian subgroup of S of maximal order. As $z \in Z(S)$, we have that $z \in A$ and in particular $z \in A \cap B$. Lemma 8.2 (1) implies that $C_A(B) \leq B$ whence $C_A(B) = A \cap B$. As $r_2(C_G(B)) = 2$, it follows that either $B = A$ or $|C_A(B)| = 2$. In the first case (4) holds. In the second case $C_A(B) = \langle z \rangle$. Our choice of B as a normal subgroup of S yields that A normalises B and that either A coincides with B or A acts non-trivially on B. Thus if $A \neq B$, then $|A : C_A(B)| \leq 2$ which leads to $|A| \leq 2 \cdot |C_A(B)| = 4$.

We conclude that $r_2(G) = 2$. $\square$

LEMMA 8.4. *Suppose that Hypothesis 8.1 holds, that $v \in \{a, b\}$ and that $C_G(v) \neq H_v$. Then $E(H_v) = 1 = O_2(H_v)$. If $C \neq M$, then $E(M) = 1 = O_2(M)$.*

PROOF. First z, v and vz are isolated in H_v by Lemma 8.2 (3). Therefore Lemma 4.2 (5) and (6) yield that $O_2(H_v)E(H_v)$ is centralised by V, but that $O_2(H_v)$ does not contain v. Also $z \notin O_2(H_v)$ or else $C_G(v) \leq M$ by Lemma 4.2 (1), contradicting Lemma 4.13.

Let us assume that $O_2(H_v) \neq 1$. Then $O_2(H_v)$ contains an involution and as $r_2(G) = 2$ by Lemma 8.3 (4), we conclude from the previous paragraph that vz is the unique involution in $O_2(H_v)$. Thus $vz \in Z(H_v)$. It follows that $C_K(v) \subseteq H_v \leq C_G(vz)$ whence H_{vz} contains $C_K(v)$ and $C_K(vz)$, contrary to Lemma 8.3 (2). Thus $O_2(H_v) = 1$.

We recall that V centralises $E(H_v)$. Then $V \cap E(H_v) \leq O_2(H_v) = 1$ and, as $r_2(G) = 2$ by Lemma 8.3 (4), this implies that $E(H_v)$ does not contain any involutions. But components have even order, so this forces $E(H_v) = 1$.

Now we turn to the case where $C < M$. With Hypothesis 8.1 let $p \in \pi$ be an odd prime such that $U \leq O_p(M)$. Assume that $O_2(M)E(M) \neq 1$ and let $t \in O_2(M)E(M)$ be an involution. Then $t \neq z$ because $O_p(M)$ is not centralised by z. Moreover $F^*(M)$ is not a p-group in this case. This means that M is as in Hypothesis 7.1 and hence Lemma 6.10 gives that $C_G(t) \leq M$. This contradicts Lemma 4.13 and forces $O_2(M)E(M) = 1$. □

LEMMA 8.5. *Suppose that Hypothesis 8.1 holds, let q be an odd prime and let $Q_1, Q_2 \in \mathcal{W}_G^*(V, q)$ be such that $Q_1 \cap Q_2 \neq 1$. Then Q_1 and Q_2 are conjugate by an element from $C_G(V)$.*

PROOF. Assume that this is not the case and choose Q_1 and Q_2 such that they are not conjugate under $C_G(V)$ and moreover such that $D := Q_1 \cap Q_2 \neq 1$ is maximal. With Lemma 8.3 (1) we know that $N_G(D)$ is a proper subgroup of G and we set $H := N_G(D)$. Then D, $N_{Q_1}(D)$ and $N_{Q_2}(D)$ are V-invariant subgroups of H. For all $i \in \{1, 2\}$ we choose $N_{Q_i}(D) \leq P_i \in \mathcal{W}_H^*(V, q)$. As q is odd and $V \leq O_{2',2}(H)$ by Lemma 4.2 (3), we may apply Lemma 2.10 which yields an element $h \in C_H(V)$ such that $P_1^h = P_2$. Now let $P_1 \leq P_1^* \in \mathcal{W}_G^*(V, q)$. Then $P_2 = P_1^h \leq (P_1^*)^h \in \mathcal{W}_G^*(V, q)$. Therefore we have that $D < N_{Q_1}(D) \leq Q_1 \cap P_1^*$ and $D < N_{Q_2}(D) \leq Q_2 \cap (P_1^*)^h$. By our choice of Q_1 and Q_2, it follows that Q_1 and P_1^* as well as Q_2 and $(P_1^*)^h$ are conjugate by an element from $C_G(V)$, respectively. We chose $h \in C_H(V)$ and therefore Q_1 and Q_2 are conjugate by an element from $C_G(V)$. This is a contradiction. □

LEMMA 8.6. *Suppose that Hypothesis 8.1 holds and that q is an odd prime such that $\text{char}(H_a) = q = \text{char}(H_b)$. Let $v \in \{a, b\}$ and suppose that Q is a V-invariant q-subgroup of G, containing $O_q(H_v)$, such that $ZJ(Q)$ is invariant under $C_K(v)$ and such that Q is maximal (with respect to inclusion) subject to these constraints. Then $Q \in \text{Syl}_q(G)$.*

PROOF. From Lemma 8.3 (1) we know that $N_G(ZJ(Q))$ is a proper subgroup of G, so let $N_G(ZJ(Q)) \leq H \max G$. Then H contains V, $C_K(v)$ and Q and $C_K(v) \subseteq O(H)$ by Lemma 4.2 (2).

Now we have that $U_v \leq O_q(H_v) \leq Q \leq H$ and Lemmas 8.2 (2) and 6.8 (3) yield that $U_v \leq F(H)$. Then Lemma 6.9 is applicable and gives that $N_G(U_v) \leq H_v$ and hence $H \looparrowright H_v$. We just saw that $U_v \leq O_q(H)$ and hence the Infection Theorem (2) implies that H has characteristic q.

With Lemma 3.11, applied to V and H_v, we choose $Q \leq Q_1 \in \mathrm{Syl}_q(H_v, V)$. It follows from $\mathrm{char}(H) = q$ that $\mathrm{char}(Q_1 O(H)) = q$ and therefore Theorem 2.23 yields that $ZJ(Q_1) \trianglelefteq Q_1 O(H) V$. Hence $ZJ(Q_1)$ is $C_K(v)$-invariant and the choice of Q gives that $Q = Q_1 \in \mathrm{Syl}_q(H)$. But $N_G(Q) \leq N_G(ZJ(Q)) \leq H$ and consequently $Q \in \mathrm{Syl}_q(N_G(Q))$. We deduce that Q is in fact a V-invariant Sylow q-subgroup of G. □

LEMMA 8.7. *Suppose that Hypothesis 8.1 holds. Then there does not exist a prime q such that $\mathrm{char}(H_a) = q = \mathrm{char}(H_b)$.*

PROOF. Assume that the result is false and that q is a prime such that H_a and H_b are both of characteristic q.

(1) $a \notin Z(H_a)$ and $b \notin Z(H_b)$. In particular q is odd.

PROOF. Let $v \in \{a, b\}$. If $v \in Z(H_v)$, then $O_2(H_v) \neq 1$ and hence $q = 2$. Thus it suffices to prove that q is odd. If $q = 2$, then $O_2(H_v)$ contains its centraliser in H_v and hence $z \in O_2(H_v)$, by Lemma 4.2 (5). Together with Lemma 4.2 (6) this forces $z \in Z(H_v)$, contradicting Lemma 4.13. □

(2) There exist subgroups $Q_1 \in \mathrm{Syl}_q(H_a, V)$ and $Q_2 \in \mathrm{Syl}_q(H_b, V)$ such that $Q_1 \cap Q_2 \neq 1$.

PROOF. Assume otherwise. With Lemma 3.11, applied to V and H_a, there exists $Q_1 \in \mathrm{Syl}_q(H_a, V)$. If $Q_1 \cap H_b \neq 1$, then this intersection is a V-invariant q-subgroup of H_b and then Lemma 3.11 implies that it is contained in some $Q_2 \in \mathrm{Syl}_q(H_b, V)$, contrary to our assumption. Therefore $Q_1 \cap H_b = 1$ and in particular Q_1 is inverted by b and hence abelian. It follows that
$$Q_1 \leq C_{H_a}(O_q(H_a)) = Z(O_q(H_a))$$
because $O_q(H_a) = F^*(H_a)$.

This implies that $Q_1 = O_q(H_a)$ and forces $N_G(Q_1)$ to be contained in H_a, because H_a is primitive (Corollary 4.8). Therefore Q_1 is a Sylow q-subgroup of G. Now the Sylow q-subgroups of G are abelian, in particular $O_q(M)$ is abelian. We recall that $O_q(H_a)$ contains an a-minimal subgroup U_a that is now abelian and (hence) inverted by a. As $O_q(H_a)$ is also inverted by b, it follows that z centralises U_a. Lemmas 4.2 (3) and 2.9 imply that $U_a = [U_a, a] \leq O_q(M)$. Now we use that $O_q(M)$ is abelian; this forces $O_q(M) \leq C_G(U_a) \leq H_a$ by Lemma 6.9 and consequently M infects H_a. The Infection Theorem (2) tells us that $\mathrm{char}(M) = q$. Moreover $O_q(M)$ lies in some z-invariant Sylow q-subgroup of G, hence in all of them by Lemma 3.10 and therefore $O_q(M) \leq Q_1$. As Q_1 is abelian, we deduce that $Q_1 \leq C_M(F^*(M)) \leq O_q(M)$. This forces $F^*(H_a) = F^*(M)$ and then $M = H_a$ by Corollary 4.8. But this is impossible by Lemma 4.13. □

With (2) we choose $Q_1 \in \mathrm{Syl}_q(H_a, V)$ and $Q_2 \in \mathrm{Syl}_q(H_b, V)$ such that $Q_1 \cap Q_2 \neq 1$. Let $v \in \{a, b\}$ and let Q_v be a V-invariant q-subgroup of G containing $O_q(H_v)$, such that $ZJ(Q_v)$ is $C_K(v)$-invariant and chosen to be maximal subject to these constraints. Then Lemma 8.6 implies that Q_v is a V-invariant Sylow q-subgroup of

G. Denoting the subgroups that we found in this way by Q_a and Q_b, we may now suppose that $Q_1 \le Q_a$ and similarly $Q_2 \le Q_b$.

As $1 \ne Q_1 \cap Q_2 \le Q_a \cap Q_b$ by our choice of Q_1 and Q_2, Lemma 8.5 yields an element $x \in C_G(V)$ such that $Q_a^x = Q_b$. Then $ZJ(Q_a)^x = ZJ(Q_b)$ and therefore $N_G(ZJ(Q_a))^x = N_G(ZJ(Q_b))$. But $C_K(a) \subseteq N_G(ZJ(Q_a))$ and this means that

$$C_K(a) = C_K(a)^x \subseteq N_G(ZJ(Q_a))^x = N_G(ZJ(Q_b)).$$

As $C_K(b) \subseteq N_G(ZJ(Q_b))$, Lemma 8.3 (2) implies that $N_G(ZJ(Q_b)) = G$. This contradicts Lemma 8.3 (1) and hence the proof is complete. □

LEMMA 8.8. *Suppose that Hypothesis 8.1 holds. Then there does not exist a prime q such that $\mathrm{char}(M) = q = \mathrm{char}(H_a)$.*

PROOF. Assume otherwise and let q be such a prime. By Lemma 8.2 (2) there are two cases to consider:

Case 1: a and b are conjugate in G.

Then $\mathrm{char}(H_b) = q$ by our choice of H_a and H_b in Hypothesis 8.1 and this contradicts Lemma 8.7.

Case 2: a and b are isolated in G.

Then the roles of a, b and z can be interchanged which makes Lemma 8.7 applicable to a and z directly. This leads to a contradiction again.

□

Before we embark on one of the main results of this section, we show that if a and b are isolated in G, then an even stronger version of Theorem **A** holds for them (and similarly for z.)

LEMMA 8.9. *Suppose that Hypothesis 8.1 holds and let $v \in \{a, b\}$. If v is isolated in G, then $H_v = C_G(v)$ or H_v has odd prime characteristic.*

PROOF. Assume that $C_G(v) < H_v$, but that H_v does not have odd prime characteristic. As v is isolated in G and $v \notin Z^*(G)$, we see that G, v, $C_G(v)$ and H_v satisfy Hypothesis 4.1 in the roles of G, z, C and M. Therefore we may apply Theorem **A** to v and H_v instead of z and M. We know from Lemma 8.4 that $E(H_v) = 1$ and hence the theorem supplies a contradiction. □

Here comes our main result:

THEOREM 8.10. *Suppose that Hypothesis 8.1 holds and let $v \in \{a, b\}$. Then $C_G(v)$ is a maximal subgroup of G and a and b are isolated in G.*

PROOF. First we assume that $C_G(v) \ne H_v$ and we set $w := vz$. Let $F := F(H_v)$ and note that $F = F_{2'}(H_v) = F^*(H_v)$ by Lemma 8.4 and that by Hypothesis 8.1 we have a v-minimal subgroup U_v in F. In particular Hypothesis 6.6 is satisfied.

(1) $[C_F(z), v] \ne 1$.

PROOF. Assume otherwise. Then $C_F(z) \le C_F(v)$. From Lemma 2.1 (4) it follows that $[F, v] \le [F, z] \cap C_G(w)$ and thus $U_v \le H_w$. As v is isolated in H_w by Lemma 8.2 (3), the Pushing Down Lemma (3) gives that $U_v \le F(H_w)$. Then Lemma 6.9 implies that $H_w \looparrowright H_v$. Moreover, Lemma 8.3 (2) and Lemma 8.7 yield that H_v and H_w are neither equal nor of the same prime characteristic. So with the Infection Theorem (5) we deduce that H_v and H_w are not conjugate. Therefore a and b are not conjugate (following from our choices in Hypothesis 8.1). Lemma 8.2 (2) yields that a and b are isolated in G and then by Lemma 8.9 there exists an odd prime q such that $\text{char}(H_v) = q$. Now the Infection Theorem (2) and the fact that $1 \neq U_v \leq O_q(H_w)$ imply that also $\text{char}(H_w) = q$. This contradicts Lemma 8.7. □

By (1) we may choose a prime $p \in \pi_v$ such that, with $P := O_p(H_v)$, we have that $X := [C_P(z), v] \neq 1$. In particular $[P, v] \neq 1$ so that we may suppose that $U_v \leq P$. As $v \in O_{2',2}(M)$ by Hypothesis 8.1, Lemma 2.9 yields that

$$X = [X, v] \leq [P \cap M, v] \leq O_p(M)$$

and therefore $X \leq C_{F(M)}(z)$.

(2) $C = M$.

PROOF. Assume that $C < M$. Then with Lemma 8.4 we know that $E(M) = 1 = O_2(M)$ and thus M has odd prime characteristic by Theorem **A**. We observed above that $1 \neq X \leq O_p(M)$ and this implies that $\text{char}(M) = p$.

As Hypothesis 6.6 is satisfied by z and M, Lemma 6.7 yields that $[X, U] \leq [C_{F(M)}(z), U] = 1$. Therefore X is a non-trivial $U\langle z \rangle$-invariant subgroup of $O_p(M)$. With Lemma 6.10 we obtain that $N_G(X)$ lies in a maximal subgroup of G of characteristic p. We also know that $X \leq F(H_v)$ whence H_v infects a maximal subgroup of G of characteristic p. But then, applying the Infection Theorem (2), we obtain that $\text{char}(H_v) = p$ contradicting Lemma 8.8. □

Now (2) and Lemma 4.10 imply that every z-invariant π-subgroup of G is contained in $C = M$. In particular we know that $[F_\pi(H_v), z] = 1$. As $X \leq O_p(M)$ and therefore $p \in \pi$, the z-invariant p-subgroup P of H_v is now contained in C. This means that $X = [P, v]$ and hence $U_v \leq X = [P, v] \trianglelefteq F$ is normalised by $U_v \langle v \rangle$.

(3) M infects H_v and (therefore) $F_{\pi'}(H_v)$ is a non-trivial subgroup that is inverted by z.

PROOF. As $U_v \leq X \leq O_p(M)$, the first statement follows from Lemma 6.9. Moreover we know that M and H_v are neither equal nor both of characteristic p (by Lemmas 4.13 and 8.8). As $E(H_v) = 1$ with Lemma 8.4, the Infection Theorem (4) gives that $F_{\pi'}(H_v) \neq 1$. This subgroup is inverted by z by the Infection Theorem (1). □

(4) H_v is the unique maximal subgroup of G containing $N_G(X)$.

PROOF. We know that $N_G(X) \neq G$ by Lemma 8.3 (1). Suppose that H is a maximal subgroup of G containing $N_G(X)$. Then as $X \leq F(H_v)$, we have that H_v infects H. Lemmas 4.2 (3) and 2.9 imply that $U_v \leq F(H)$ and therefore Lemma 6.9 yields that we conversely have that $H \looparrowright H_v$. Now the Infection Theorem (3) forces H and H_v to be equal or both of characteristic p. But if $\text{char}(H) = p$, then we recall that $X \leq O_p(M)$ and hence $M \looparrowright H$. Then the Infection Theorem (2) implies that $\text{char}(M) = p$ as well, contrary to Lemma 8.8. Thus $H = H_v$ as stated. □

(5) a and b are conjugate in G.

PROOF. Otherwise Lemma 8.2 (2) implies that a and b are isolated in G and hence that v is isolated in G. Together with our assumption that $C_G(v) < H_v$, Lemma 8.9 implies that $\mathrm{char}(H_v) = p$. But now, since $F^*(H_v) = P \leq M$, it follows that $H_v \looparrowright M$. As $M \looparrowright H_v$ by (3), the Infection Theorem (3) and Lemma 8.8 imply that $M = H_v$. This contradicts Lemma 4.13. $\square$

By Hypothesis 8.1 we now have that H_a and H_b are conjugate and in particular it follows that $C_G(w) < H_w$ and that $Y := [O_p(H_w), w] \neq 1$. We also recall that $F_\pi(H_v)$ is centralised by z (as noted before (3)) and that $F_{\pi'}(H_v)$ is inverted by z by (3). In particular $F_{\pi'}(H_v)$ is abelian. Then we see that $[H_v, z] \leq C_{H_v}(F) \leq F$, because $F = F^*(H_v)$, and together with Lemma 2.5 this implies that $H_v = C_{H_v}(z) F_{\pi'}(H_v)$. We point out that, although a and b are conjugate in G, they are still isolated in H_a and in H_b, by Lemma 8.2 (3).

(6) $Y \leq O_p(M)$ and $N_G(Y) \leq H_w$. In particular M infects H_w.

PROOF. We know that v and w are conjugate in G by (5) and hence in C by Lemma 3.1 (9). Then X and Y are conjugate by an element from C and the first statement follows because $X \leq O_p(M)$. Again by conjugacy and by (4), the unique maximal subgroup of G containing $N_G(Y)$ is H_w. Thus $M \looparrowright H_w$. $\square$

(7) $F_{\pi'}(H_v)$ is inverted by w and by z and (therefore) centralised by v.

PROOF. Let $D := F_{\pi'}(H_v) \cap C_G(w)$. Then $D \trianglelefteq F_{\pi'}(H_v)$ because $F_{\pi'}(H_v)$ is abelian. Moreover D is invariant under $C_G(w) \cap C_G(v) = C_C(v) = C_C(w)$. Since $[C_G(v), z] \leq [H_v, z] \leq F_{\pi'}(H_v)$ (as noted before (6)), Lemma 4.2 (1) gives that

$$C_G(v) = [C_G(v), z] C_C(v) \leq F_{\pi'}(H_v) C_C(v).$$

Consequently $C_G(v)$ normalises D. Moreover D is contained in $C_G(w)$ and inverted by z, so it follows that

$$D = [D, z] \leq [C_G(w), z] \leq [H_w, z] \leq F_{\pi'}(H_w).$$

As v and w are conjugate in G by (5), the subgroup $F_{\pi'}(H_w)$ is abelian as well. In particular $D \trianglelefteq F_{\pi'}(H_w)$. Now we can argue as above to deduce that D is also $C_G(w)$-invariant. Together with Lemma 8.3 (2) we see that $G = \langle C_G(v), C_G(w) \rangle \leq N_G(D)$. This forces $D = 1$ because G is simple by part (1) of the same lemma. Hence w inverts $F_{\pi'}(H_v)$. This subgroup is also inverted by z, by (3), and hence v centralises it. $\square$

(8) $O_p(M) \not\leq H_v$.

PROOF. Otherwise $Y \leq O_p(M) \leq H_v$ and it follows that $Y \leq O_p(H_v)$ with the Pushing Down Lemma (1). Then $H_v \looparrowright H_w$ by (6). As $E(H_v) = 1 = E(H_w)$ by Lemma 8.4 and as v and w are conjugate by (5), the Infection Theorem (5) forces H_v and H_w to be equal or to be of characteristic p. But this is contradicted by Lemma 8.7 and the fact that H_v and H_w are distinct (by Lemma 8.3 (2)). $\square$

(9) $\pi \cap \pi_v = \{p\}$.

PROOF. Assume that there exists a prime $q \in \pi \cap \pi_v$ such that $q \neq p$. Then, as z centralises $F_\pi(H_v)$, we have that $P \times O_q(H_v)$ acts on $O_p(M)$ and
$$[C_{O_p(M)}(P), O_q(H_v)] \leq O_p(M) \cap O_q(H_v) = 1.$$
With Thompson's $P \times Q$-Lemma (2.2) it follows that $[O_p(M), O_q(H_v)] = 1$ and therefore $O_p(M) \leq H_v$, contradicting (8). □

(10) $X = P \cap O_p(M)$. Moreover $X \trianglelefteq H_v$ and X is a cyclic group that is inverted by v and w.

PROOF. Let $P_0 := P \cap O_p(M)$.

First assume that $Z := \Omega_1(Z(O_p(M))) \leq P_0$. Then $Z \leq P$ and consequently $F_{\pi'}(H_v) \leq C_G(Z) \leq M$ (because M is primitive by Corollary 4.8). This contradicts (2) and (3). Thus $Z \not\leq P_0$.

Now assume that P_0 is not cyclic. Then choose $Q_0 \leq P_0$ to be elementary abelian of order p^2 and let $Q_0 \leq Q_1 \leq Q_0 Z$ be such that Q_1 is elementary abelian of order p^3. This choice is possible by the previous paragraph. As $M = C$ by (2), we may apply Theorem 7.9 and we obtain that $C_G(x) \leq M$ for all $x \in Q_1^\#$. In particular $C_G(Q_0) \leq M$. On the other hand $Q_0 \leq P$ and thus
$$F_{\pi'}(H_v) \leq F_{\pi'}(H_v) \leq C_G(Q_0) \leq M,$$
which is impossible by (2) and (3). We conclude that P_0 is cyclic and hence P_0 is either centralised or inverted by v. Now we recall that $U_v \leq X \leq P_0$. Then $[P_0, v] \neq 1$, therefore v inverts P_0. Moreover P_0 is centralised by z and hence inverted by w. It follows that $X \leq P_0 = [P_0, v] \leq [P, v] = X$, so $X = P_0$ as stated.

It remains to show that $X \trianglelefteq H_v$. Of course $F_{\pi'}(H_v)$ centralises X because $p \in \pi$, moreover $X = P \cap O_p(M)$ is $C_{H_v}(z)$-invariant. Thus the fact that $H_v = C_{H_v}(z) F_{\pi'}(H_v)$ yields that X is normal in H_v. □

(11) For all primes $q \neq p$, there exists a Sylow q-subgroup of H_v that is centralised by v.

PROOF. As X is inverted by v, by (10), and as X is normal in H_v, we have that $[H_v, v]$ centralises X. Then we recall that $P = C_P(v)[P, v] = C_P(v)X$ by Lemma 2.1 (2) and we consider the action of $[H_v, v]$ on P/X. It follows that P/X is centralised by $[H_v, v]$ as well. Moreover (7) and (9) yield that $[F_{p'}(H_v), v] = 1$ and therefore $[H_v, v]$ centralises $F_{p'}(H_v)$. This implies that $[H_v, v] \leq O_p(H_v)C_{H_v}(F^*(H_v)) = PZ(F)$. Since $[H_v, v, v] = [H_v, v]$ by Lemma 4.2 (1) and v centralises $F_{p'}(H_v)$, we deduce that
$$[H_v, v] \leq [PZ(F), v] = [P, v] = X.$$
Suppose that $q \neq p$ is a prime and, with Lemma 3.11, let $Q \in \mathrm{Syl}_q(H_v, V)$. (Here we use again that all involutions in V are isolated in H_v, by Lemma 8.2 (2).) Then it follows that
$$[Q, v] \leq [H_v, v] \cap Q \leq X \cap Q = 1.$$
□

With (3) we find a prime $q \in \pi' \cap \pi_v$. Then $q \neq p$ and, by (11), we may choose $Q \in \mathrm{Syl}_q(H_v, V)$ such that v centralises Q. Moreover (3) and (9) give that $F_{p'}(H_v) = F_{\pi'}(H_v)$ is non-trivial and is inverted by z, so $O_q(H_v)$ is inverted by z and therefore Q is not centralised by z. As $C_G(Q)$ is z-invariant, we may choose $v \in T \in \mathrm{Syl}_2(C_G(Q))$ such that T is z-invariant (by Lemma 3.8, applied to $C_G(Q)\langle z \rangle$). Then $[T, z] = 1$ by Lemma 3.1 (2), but $z \notin T$. Together with Lemma 8.3 (4) it follows that v is the unique involution in T and therefore $N_G(T) \leq C_G(v)$. However, we know from (5) that Lemma 6.11 (1) holds and hence that v is not the only involution in $Z(T)$. This contradiction shows that $C_G(v)$ is a maximal subgroup of G as stated. It is left to prove that a and b are isolated in G:

As z is not central in H_a by Lemma 4.13, it follows from Lemma 4.2 (6) that there exists an odd prime $q \in \pi_a$ such that $O_q(H_a)$ is not centralised by z. Lemma 8.2 (2) gives that all involutions in V are isolated in H_a, so with Lemma 3.11 we may choose $Q \in \mathrm{Syl}_q(H_a, V)$ and $a \in T \in \mathrm{Syl}_2(C_G(Q))$ such that T is z-invariant, as we did in the previous paragraph. Then z centralises T, but is not contained in it. Again the fact that $r_2(G) = 2$ by Lemma 8.3 (4) yields that a is the unique involution in T. Thus Lemma 6.11(2) must hold and in particular a and b are not conjugate in G. Lemma 8.2 (2) implies that a and b are isolated in G.

This finishes the proof of the theorem. $\square$

COROLLARY 8.11. *Suppose that Hypothesis 8.1 holds. Then $C = M$.*

PROOF. We know from Theorem 8.10 that a and b are isolated in G. Thus the roles of z, a and b can be interchanged. In particular, since $z \in O_{2',2}(C) \cap C_G(a) \leq O_{2',2}(C_G(a))$ by Lemma 4.2 (3), we may apply Theorem 8.10 to z instead of v. Hence C is a maximal subgroup of G. $\square$

8.2. The Proof of Theorem B

After a preparatory lemma, we set up a hypothesis following the results in the previous part of this chapter. Then we apply local analysis to involution centralisers and we arrive at a contradiction with a counting argument. Hypothesis 4.1 is assumed to hold throughout.

HYPOTHESIS 8.12.
In addition to Hypothesis 4.1, suppose that $a \in O_{2',2}(C)$ is an involution distinct from z. Moreover

- $V := \langle a, z \rangle$ and $b := az$;
- *for all $v \in \{a, b\}$ we let $H_v := C_G(v)$ and $\pi_v := \pi(F(H_v))$ and*
- $\pi := \pi(F(C))$.

LEMMA 8.13. *Suppose that Hypothesis 8.12 holds and let $V \leq S \in Syl_2(G)$. Then Hypothesis 8.1 is satisfied, the subgroups C, H_a and H_b are maximal in G and in particular, for all $v \in V^\#$, Hypothesis 6.6 holds with $v, C_G(v)$ and H_v in the roles of $t, C_G(t)$ and H_t. In S there are precisely three involutions and these are central in S and isolated in G.*

PROOF. Hypothesis 8.12 together with some more notation and choices of maximal subgroups gives Hypothesis 8.1. If $v \in V^\#$, then Lemma 8.2 (3) implies that v, $C_G(v)$ and H_v satisfy Hypothesis 6.6. Theorem 8.10 and Corollary 8.11 are applicable and hence C, H_a and H_b are maximal subgroups and a and b are isolated in G. Thus all involutions in V are isolated in G and hence central in S. Lemma 8.3 (4) yields that $r_2(G) = 2$ and therefore $\Omega_1(S) = V$. □

LEMMA 8.14. *Suppose that Hypothesis 8.12 holds and suppose further that* $[C, a] \not\leq F(C)$. *Then* $[H_a, z] \leq F(H_a)$ *and* $[H_b, z] \leq F(H_b)$.

PROOF. By Lemma 8.13 we may apply Lemma 6.12 with a and z interchanged. Then it follows that $O(F(H_a)) \cap C = 1$ and hence that z inverts $O(F(H_a))$. We also know from Lemma 4.2 (5) that z centralises $O_2(H_a)E(H_a)$. Therefore $F^*(H_a)$ is centralised by $[H_a, z]$ and we deduce that $[H_a, z] \leq C_{H_a}(F^*(H_a)) = Z(F(H_a))$. In the same way we argue that $[H_b, z] \leq F(H_b)$. □

LEMMA 8.15. *Suppose that Hypothesis 8.12 holds. Suppose that p is a prime and that P is a V-invariant Sylow p-subgroup of G. Then P is centralised by a, b or z.*

PROOF. If $p \in \pi$, then Lemmas 8.13 and 4.10 give that P is centralised by z. Thus we now suppose that $[P, z] \neq 1$ and hence that $p \notin \pi$. Then Lemma 3.7 implies that p divides $|C_K(a)|$ or $|C_K(b)|$. In the first case, the same lemma implies that $|H_a|$ is divisible by p, but that $C_{H_a}(z)$ does not contain a Sylow p-subgroup of H_a. Therefore $|[H_a, z]|$ is divisible by p. If $[H_a, z] \not\leq F(H_a)$, then Lemma 8.14, with a in the role of z, yields that $[C, a] \leq F(C)$ and $[H_b, a] \leq F(H_b)$. (Here we use that a and b are isolated in G by Lemma 8.13.) In particular, since $p \notin \pi$, we see that a centralises every V-invariant p-subgroup of C.

If, in this case, we have that $p \in \pi_b$, then Lemma 4.10, applied to b, gives that H_b contains a Sylow p-subgroup of G. Lemma 3.10 then implies that $P \leq C_G(b)$.

If, still in the same case, we have that $p \notin \pi_b$, then a also centralises every V-invariant p-subgroup of H_b. In particular $[C_P(z), a] = 1 = [C_P(b), a]$ because P is V-invariant. Then by Lemma 2.1 (4) it follows that $[P, a] = 1$.

Finally, if $[H_a, z] \leq F(H_a)$, then $p \in \pi(F(H_a))$ and Lemmas 4.10 and 3.10, applied to a, yield that $[P, a] = 1$. □

LEMMA 8.16. *Suppose that Hypothesis 8.12 holds. Then $\overline{C}$ is perfect.*

PROOF. Assume otherwise. Then $\overline{C}' < \overline{C}$ which means that $\overline{C}$ possesses a non-trivial abelian factor group. As $O(C_G(V)) \leq O(C)$ by Lemma 2.8, we have that $O(C_G(V)) = O(C) \cap C_G(V)$ and therefore $\overline{C} \simeq C_G(V)/O(C_G(V))$. Now $C_G(V)/O(C_G(V))$ possesses a non-trivial abelian factor group. We recall that a and b are isolated in G by Lemma 8.13. Therefore the same arguments as above give that
$$H_a/O(H_a) \simeq C_G(V)/O(C_G(V)) \simeq H_b/O(H_b).$$
Hence there exists some prime p such that $\overline{C}$, $H_a/O(H_a)$ and $H_b/O(H_b)$ have a non-trivial p-factor group. But for the same prime p, there exists a V-invariant Sylow p-subgroup P of G by Lemma 3.11. Then with Lemma 8.15 it follows that some $H \in \{H_a, H_b, C\}$ contains P. As Hypothesis 8.1 is satisfied by Lemma 8.13,

Lemma 8.3 (1) yields that G is simple and hence $O^p(G) = G$. Then Lemmas 3.1 (3) and 2.18 imply that H does not have any non-trivial p-factor group, which is a contradiction. □

LEMMA 8.17. *Suppose that Hypothesis 8.12 holds and let $p \in \pi$ be such that $O_p(C)$ is not centralised by a. Then $C_{O_p(C)}(a)$ is cyclic.*

PROOF. Lemma 8.13 implies that $[O_2(C), a] = 1$ and therefore p is odd. Set $P := O_p(C)$ and assume that $C_P(a)$ contains an elementary abelian subgroup X of order p^2. If X is a maximal elementary abelian subgroup of P, then $X = \Omega_1(C_P(X))$ and therefore a centralises every element of order p in $C_P(X)$. As p is odd, it follows with Lemma 2.1 (5) and (6) first that a centralises $C_P(X)$ and then that a centralises P. This is a contradiction. Thus X lies in an elementary abelian subgroup Y of P of order p^3. Theorem 7.9 forces $C_G(y)$ to be contained in C for all $y \in Y^\#$. But we have that $X \leq C_P(a) \leq H_a$ and therefore X acts coprimely on $O_{p'}(H_a)$. With Lemma 2.1 (4) we obtain that

$$O_{p'}(H_a) = \langle C_{O_{p'}(H_a)}(x) \mid x \in X^\# \rangle \leq C.$$

As $O_p(H_a)$ is a z-invariant p-subgroup and $p \in \pi$, Lemma 4.10 implies that $O_p(H_a) \leq C$. Therefore z centralises $F(H_a)$. But this contradicts Lemmas 4.13 and 4.2 (6). □

LEMMA 8.18. *Suppose that Hypothesis 8.12 holds and let $v, w \in V^\#$. Then $[C_G(v), w] \leq F(C_G(v))$.*

PROOF. By Lemma 8.13, all involutions in G are isolated in G. Therefore it is sufficient to prove that $[C, a] \leq F(C)$. As a is isolated in G, Lemma 4.2 (5) gives that $[a, O_2(C)E(C)] = 1$. It is left to show that, for all odd primes $p \in \pi$, the subgroup $[C, a]$ acts nilpotently on $O_p(C)$. Thus let $p \in \pi$ be odd. If $[O_p(C), a] = 1$, then nothing is left to prove. If $O_p(C)$ is not centralised by a, then Lemma 8.17 implies that $C_{O_p(C)}(a)$ is cyclic. From Lemma 8.16 we know that $C/O(C)$ is 2- and 3-perfect and thus Corollary 2.30 is applicable. It yields that $[C, a]$ acts nilpotently on $O_p(C)$ as required. □

COROLLARY 8.19. *Suppose that Hypothesis 8.12 holds. Suppose that p is a prime in $\pi(G) \setminus \pi \cup \pi_a \cup \pi_b$ and that $P \in Syl_p(G)$ is V-invariant. Then P is centralised by V.*

PROOF. From Lemma 8.15 we know that P is centralised by some involution in V. Since they are all isolated, we may without loss suppose that $P \leq C$. Then Lemma 8.18 yields that $[P, a] \leq F(C)$. The choice of p implies that p does not divide $|F(C)|$ and hence $[P, a] = 1$. It follows that P is centralised by V. □

LEMMA 8.20. *Suppose that Hypothesis 8.12 holds and suppose that $x \in F(C)$ is an element that is inverted and not centralised by a. Then C is the unique maximal subgroup of G containing $C_{F^*(C)}(x)\langle a \rangle$. In particular the only maximal subgroup of G containing $C_G(x)\langle a \rangle$ is C.*

PROOF. As a centralises $O_2(C)$, our hypothesis forces $x \in O(F(C))$. Therefore we may suppose that x is a p-element for some odd prime $p \in \pi$. In particular $|\pi| \geq 2$.

Set $Y_0 := C_{F(C)}(x)$ and $Y := Y_0 E(C)$. Then Y is centraliser closed in $F(C)$. Suppose that $Y\langle a \rangle \leq H \max G$ and assume that $H \neq C$. Then $z \in H$, moreover C infects H and $E(H) \leq C$ by Lemma 4.2 (5). With the Infection Theorem (4) there exists a prime $q \in \pi'$ such that $Q := O_q(H) \neq 1$. Then Lemma 7.8 says that Q is the unique maximal Y-invariant q-subgroup of G intersecting H non-trivially. Moreover z inverts Q by Lemma 7.7 (3).

(1) Q is centralised by a or by b.

PROOF. Q is a V-invariant q-group because Q is normal in H and $V \leq H$. Let $Q \leq Q_1 \in \mathcal{W}_q^*(G, V)$. As z inverts Q, it follows that Q_1 is not centralised by z, thus with Lemma 8.15 it is centralised by a or by b. □

By symmetry between a and b, we suppose that $Q \leq H_a$.

(2) $p < q$.

PROOF. We know that $Q = [Q, z] \leq F(H_a)$ by Lemma 8.18. As H is primitive by Corollary 4.8, we have that $N_G(Q) = H$ and consequently H_a infects H. We note that a centralises $E(H)$ by Lemma 4.2 (5).

Now assume that $r(Q) \geq 3$ (and thus $r(O_q(H_a)) \geq 3$). Then Theorem 7.9, applied to H_a, yields that $C_G(y) \leq H_a$ for all elements y of order q that lie in some elementary abelian subgroup of order q^3 of $O_q(H_a)$. With Lemma 2.1 (4) this forces $O_{q'}(H) \leq H_a$. Hence $F^*(H)$ is centralised by a. With Lemma 4.2 (6) it follows that $H = H_a$, which is impossible because $x \in H$ is inverted by a. Thus we have that $r(Q) \leq 2$. Let $Y_p := O_p(Y_0)$. Then $C_G(Y_p) \leq C$ by Lemma 7.7 (1) and hence $C_Q(Y_p) \leq C \cap Q = 1$ by part (3) of the same lemma. In particular Y_p acts non-trivially on Q. As $r(Q) \leq 2$, we deduce that $p < q$ with Lemma 2.3. □

(3) $q < p$.

PROOF. Let $Q^* := QC_{F(H_a)}(Q)$. We know that $H \neq H_a$ and that $Q^*\langle z \rangle \trianglelefteq H$ because $Q \trianglelefteq H$. As x is inverted by a, Lemma 7.7 (4) implies that $x \in O_p(H)$. In particular $O_p(H) \neq 1$. Let P be a maximal $Q^*E(H_a)$-invariant p-subgroup P of G. Then Lemma 7.7 and the previous two steps, applied to H_a instead of C, give the following:

P is inverted by a and centralised by z or by b, and $r(P) \leq 2$. Moreover $O_q(C_{F(H_a)}(Q))$ acts non-trivially on P and hence $q < p$ by Lemma 2.3. □

As (2) and (3) contradict each other, the proof is complete. □

LEMMA 8.21. *Suppose that Hypothesis 8.12 holds and let $x \in K^\#$. Then there exists a unique involution $u \in a^C \cup b^C$ that centralises x.*

8.2. THE PROOF OF THEOREM B

PROOF. The order of x is odd by Lemma 3.3 (2). Thus there exists a power y of x that is a non-trivial p-element for some odd prime p, and y lies in some z-invariant Sylow p-subgroup P of G by Lemma 3.8. Lemma 13.7 yields that Lemma 3.11 is applicable whence G possesses V-invariant Sylow p-subgroups. Then it follows with Lemma 3.10 that P is C-conjugate to some V-invariant Sylow p-subgroup of G. Let $c \in C$ be such that P is invariant under $V_1 := V^c$. With Lemma 8.15 we find an involution $u \in V_1$ that centralises P and z. We note that $u \neq z$ because z inverts y. Hence u is conjugate to a or b (even in C, by Lemma 3.1 (9)) because a, b and z are representatives for the three distinct conjugacy classes of involutions in G (see Lemma 8.13). We have that $y \in C_G(u)$ and $\langle y \rangle = [\langle y \rangle, z] \leq [C_G(u), z]$. Now Lemma 8.18, applied to $C_G(u)$ instead of C, yields that $y \in F(C_G(u))$. Lemma 8.20 implies that $C_G(u)$ is the unique maximal subgroup of G containing $C_G(y)\langle z \rangle$.

Assume that some involution $u' \in a^C \cup b^C$ distinct from u centralises x. Then $u' \in C_G(y)$ and hence $\langle y \rangle = [\langle y \rangle, z] \leq [C_G(u'), z] \leq F(C_G(u'))$, again with Lemma 8.18. Thus Lemma 8.20 forces $C_G(u')$ to be the unique maximal subgroup of G containing $C_G(y)\langle z \rangle$. This means that $C_G(u) = C_G(u')$. In particular u and u' commute, so uu' is an involution because $u \neq u'$. As u and u' are distinct from z, we deduce that $uu' \in z^G$. But uu' centralises z whence $uu' = z$, because z is isolated in G. Now we recall that u and u' centralise y. Then z centralises y, but z inverts x, and this is impossible. □

COROLLARY 8.22. *Suppose that Hypothesis* 8.12 *holds. Then*
$$|K^\#| = |a^C| \cdot |C_{K^\#}(a)| + |b^C| \cdot |C_{K^\#}(b)|.$$

PROOF. By Lemma 8.21, every element in $K^\#$ is centralised by either precisely one conjugate of a or by precisely one conjugate of b. This yields the formula. □

PROOF OF THEOREM B.

Recall that in Theorem B, we suppose that G is a minimal counter-example to the Z*-Theorem and therefore Hypothesis 4.1 is satisfied. We assume that Theorem B does not hold. Then there exists an involution $a \in O_{2',2}(C)$ distinct from z, so, together with some notation, Hypothesis 8.1 holds. Corollary 8.22 gives us three formulas when applied to the sets K, $K_a := \{aa^g \mid g \in G\}$ and $K_b := \{bb^G \mid g \in G\}$, respectively:
$$|K^\#| = |a^C| \cdot |C_{K^\#}(a)| + |b^C| \cdot |C_{K^\#}(b)|,$$
$$|K_a^\#| = |b^{H_a}| \cdot |C_{K_a^\#}(b)| + |z^{H_a}| \cdot |C_{K_a^\#}(z)|$$
and
$$|K_b^\#| = |z^{H_b}| \cdot |C_{K_b^\#}(z)| + |a^{H_b}| \cdot |C_{K_b^\#}(a)|.$$

As $|a^C| = |C : C_C(a)| = |C : C_C(b)|$, we see that

(∗) $\qquad |a^C| = |b^C|$ and similarly $|b^{H_a}| = |z^{H_a}|$ and $|z^{H_b}| = |a^{H_b}|$.

We also note that the numbers in (∗) are all odd and, by Lemma 4.13, not equal to 1. This means that they are greater than or equal to 3. Next we observe that Lemma 3.7 implies that $|H_a| = |C_C(a)| \cdot |C_K(a)|$ and therefore
$$|C_K(a)| = |H_a : C_C(a)| = |z^{H_a}|.$$

Applying this for a, b and z respectively yields
$$|C_{K^\#}(a)| = |z^{H_a}| - 1 \;,\; |C_{K^\#}(b)| = |z^{H_b}| - 1,$$
$$|C_{K_a^\#}(b)| = |a^{H_b}| - 1 \;,\; |C_{K_a^\#}(z)| = |a^C| - 1,$$
$$|C_{K_b^\#}(z)| = |b^C| - 1 \;,\; |C_{K_b^\#}(a)| = |b^{H_a}| - 1.$$

Finally we apply Theorem 3.6 to deduce that
$$|K^\#| = |C_K(a)| \cdot |C_K(b)| - 1 = |z^{H_a}| \cdot |z^{H_b}| - 1$$
and similarly, for the other two involutions, we have that
$$|K_a^\#| = |a^{H_b}| \cdot |a^C| - 1 \text{ and } |K_b^\#| = |b^C| \cdot |b^{H_a}| - 1.$$

With all this in mind, the equations above become
$$|z^{H_a}| \cdot |z^{H_b}| - 1 = |a^C| \cdot (|z^{H_a}| - 1) + |a^C| \cdot (|a^{H_b}| - 1) = |a^C| \cdot |z^{H_a}| - 2|a^C| + |a^C| \cdot |a^{H_b}|,$$
$$|a^{H_b}| \cdot |a^C| - 1 = |b^{H_a}| \cdot (|a^{H_b}| - 1) + |b^{H_a}| \cdot (|b^C| - 1) = |b^{H_a}| \cdot |a^{H_b}| - 2|b^{H_a}| + |b^{H_a}| \cdot |b^C|$$
and
$$|b^C| \cdot |b^{H_a}| - 1 = |z^{H_b}| \cdot (|b^C| - 1) + |z^{H_b}| \cdot (|z^{H_a}| - 1) = |z^{H_b}| \cdot |b^C| - 2|z^{H_b}| + |z^{H_b}| \cdot |z^{H_a}|.$$

Addition of these equations and replacing terms, referring to $(*)$, yields that
$$-3 = -2|a^C| - 2|b^{H_a}| - 2|z^{H_b}| + |a^C| \cdot |z^{H_a}| + |b^{H_a}| \cdot |a^{H_b}| + |z^{H_b}| \cdot |b^C|$$
$$= |a^C|(|z^{H_a}| - 2) + |b^{H_a}|(|a^{H_b}| - 2) + |z^{H_b}|(|b^C| - 2)$$
$$\geq |a^C| + |b^{H_a}| + |z^{H_b}| \geq 3 + 3 + 3 = 9$$
which is impossible. $\square$

CHAPTER 9

Components of $\overline{C}$ and the Soluble Z*-Theorem

In this section we use Theorem **B** to show that $E(\overline{C}) \neq 1$ and to limit the number of components. As an independent result on the way, we prove the Soluble Z*-Theorem. Throughout, we suppose that Hypothesis 4.1 holds.

THEOREM 9.1. $\mathcal{L}_2(C) \neq \varnothing$.

PROOF. Assume otherwise. Then $E(\overline{C}) = 1$ and therefore $F^*(\overline{C}) = O_2(\overline{C})$. From Theorem **B** we know that $r_2(O_{2',2}(C)) = 1$ and therefore a Sylow 2-subgroup T of $O_{2',2}(C)$ is cyclic or quaternion. In the first case $\mathrm{Aut}(T)$ is a cyclic 2-group and we deduce that $[O^2(\overline{C}), \overline{T}] = 1$. Then $O^2(\overline{C}) \leq C_{\overline{C}}(\overline{T}) \leq \overline{T}$ which means that $\overline{C}$ is a 2-group and in particular C has cyclic Sylow 2-subgroups. This contradicts Lemma 4.3 (1). In the second case, assume that T is quaternion of order at least 16. Then $\mathrm{Aut}(T)$ is a 2-group and therefore $O^2(\overline{C}) \leq C_{\overline{C}}(\overline{T}) \leq \overline{T}$ as in the previous case. Then C has quaternion Sylow 2-subgroups, again contrary to Lemma 4.3 (1).

Thus we consider the situation where $T \simeq Q_8$ and we recall that $\mathrm{Aut}(Q_8) \simeq S_4$. As T is quaternion with central involution z, it follows that $z \in T' \leq G' = F^*(G)$ by Lemma 4.4. Therefore $G = F^*(G)$ is simple. Lemma 2.18 yields that $O^2(\overline{C}) = \overline{C}$ and in particular $\overline{C}/C_{\overline{C}}(\overline{T}) = \overline{C}/Z(\overline{T})$ is isomorphic to a subgroup of A_4. Let $T \leq S \in \mathrm{Syl}_2(G)$. Then $S \leq C$ and it follows that S induces inner automorphisms on $\overline{T}$. Therefore $S = T \simeq Q_8$, which is impossible by Lemma 4.3 (1). □

The next objective is to prove Theorem **C**. This is where the notion of core-separated subgroups that we introduced in Section 5 comes into play. Also, we appeal to a result that depends on a theorem usually referred to as "L-Balance". We state this here, in an appropriate way, for our minimal counter-example G. The reason why it is not listed among the general results is that, ultimately, it depends on the Z*-Theorem, because one of the main ingredients for its original proof is Glauberman's result on automorphism groups of core-free groups (see [**Gla66b**]). However, this still means that the L-Balance Theorem holds in the class of groups that satisfy the Z*-Theorem, i.e. it holds in every proper subgroup of G by Hypothesis 4.1.

THEOREM 9.2. *Suppose that $H < G$ and that $t \in H$ is an involution. Then $L(C_H(t)) \leq L(H)$.*

PROOF. The full result is stated as Theorem 4.73 in [**Gor82**]. □

LEMMA 9.3. *Let $a \in O_{2', F^*}(C)$ be an involution and suppose that L_1 is a 2-component of $C_G(a)$. Then $O^\infty(C_{L_1}(z))$ is contained in a 2-component of C.*

PROOF. Let $H := C_G(a)$ and set $\widehat{H} := H/O(H)$. We have that $\widehat{H} = \widehat{C_H(z)}$ by Lemma 4.2 (1) because $z \in H < G$. By hypothesis $L_1 \in \mathcal{L}_2(H)$ and we set $L := \langle L_1^H \rangle$ and $L_0 := O^\infty(C_L(z))$. As L is normal in H, it follows that L_0 is normal in $C_H(z) = C_C(a)$. We also note that L_0 is perfect and that $\widehat{L_0} = O^\infty(C_{\widehat{L}}(\widehat{z})) = O^\infty(\widehat{L}) = \widehat{L}$. Now
$$L_0 \leq O^{2'}(C_L(z)) \leq L(C_C(a)) \leq L(C)$$
by Theorem 9.2 and thus $L_0 \leq C_{L(C)}(a)$. In particular, we see that $\overline{L_0} \leq C_{E(\overline{C})}(\overline{a})$ and hence that $E(\overline{C}) \neq 1$. Let $F := O_{2',F^*}(C)$, let $n \in \mathbb{N}$ and let $E_1, ..., E_n$ be the 2-components of C. We note that $E_1, ..., E_n$ are normal in F and in particular a-invariant.

We need to show that there exists some $j \in \{1, ..., n\}$ such that $O^\infty(C_{L_1}(z)) \leq E_j O(C)$. But since $C_{L_1}(z)$ is not necessarily normal in $C_C(a)$, it is more convenient to first look at L_0 a bit more.

As $a \in F$, it follows that a can be written as $a = a_1 \cdots a_n t$ with $t \in O_{2',2}(C)$ and $a_j \in E_j$ for all $j \in \{1, ..., n\}$. We let $X, X_1, ..., X_n$ be normal subgroups of $C_C(a)$ such that $\overline{X} = C_{O_2(\overline{C})}(\overline{a}) = C_{O_2(\overline{C})}(\overline{t})$ and $\overline{X_j} = C_{\overline{E_j}}(\overline{a_j}) = C_{\overline{E_j}}(\overline{a})$ for all $j \in \{1, ..., n\}$. Then $C_{\overline{F}}(\overline{a}) = \overline{X X_1} \cdots \overline{X_n}$ and $\overline{L_0}$ is a perfect, normal subgroup of $C_{E(\overline{C})}(\overline{a})$, in particular $\overline{L_0}$ centralises $\overline{X}$.

Suppose that $\overline{L_0}$ is not contained in $\overline{X_1}$. Then $X_1 \cap L_0$ is a proper normal subgroup of L_0, in particular it is subnormal in L. It follows that $\widehat{X_1 \cap L_0} = \widehat{L}$ or that $\widehat{X_1 \cap L_0}$ is either a component of $\widehat{L}$ (and hence conjugate to $\widehat{L_1}$ in $C_{\widehat{H}}(\widehat{z})$) or it is contained in $Z(\widehat{L})$. A similar statement holds for $\overline{X_2}, ..., \overline{X_n}$ if $\overline{L_0}$ is not contained in either of these subgroups. If for all $i \in \{1, ..., n\}$ we have that $\widehat{X_i \cap L_0} \leq Z(\widehat{L})$, then
$$L_0 \leq (X \cap L_0)(X_1 \cap L_0) \cdots (X_n \cap L_0)O(C)$$
because modulo $O_2(\overline{C})$, the product $\overline{X X_1} \cdots \overline{X_n}$ is direct and because $\overline{L_0}$ is perfect. Thus
$$L_0 = (X \cap L_0)(X_1 \cap L_0) \cdots (X_n \cap L_0)(L_0 \cap O(C)).$$
We recall that $L_0 \cap O(C) \leq O(C_C(a)) \leq O(H)$ by Lemma 2.8. Consequently, with our observation from the previous paragraph, $\widehat{L_0}$ is contained in $Z(\widehat{L})$ and hence is abelian, which is a contradiction.

We deduce that there exists some $i \in \{1, ..., n\}$ such that $\widehat{X_i \cap L_0}$ is a component of $\widehat{L}$ or coincides with $\widehat{L}$. If it is a component, then it is $\widehat{H}$-conjugate to $\widehat{L_1}$ and hence $C_{\widehat{H}}(\widehat{z})$-conjugate to it, by Lemma 3.1 (9). As $X_i \cap L_0$ is $C_H(z)$-invariant, it follows in both cases that
$$\widehat{L} = \widehat{L_0} = O^\infty(\widehat{C_L(z)}) = O^\infty(\widehat{C_{L_1}(z)}) \leq \widehat{X_i \cap L_0}.$$
Set $J := (X_i \cap L_0)O(H)$. Then $L_0 \leq J$ and we see that $C_L(z) \leq C_J(z) = (X_i \cap L_0)C_{O(H)}(z)$ by Lemma 2.1 (3). Moreover $X_i \cap L_0$ is normalised by $C_{O(H)}(z)$ and hence normal in $C_J(z)$. This implies that $O^\infty(C_J(z)) \leq X_i \cap L_0$. But we also have that L_0 is normal in $C_J(z)$ and hence that
$$O^\infty(C_{L_1}(z)) \leq O^\infty(L_0) \leq O^\infty(C_J(z)) \leq X_i \leq E_i.$$

Therefore, in this case, our statement is proved. Now we may suppose that there exists some $j \in \{1, ..., n\}$ such that $\overline{L_0} \leq \overline{X_j}$ and hence $L_0 O(C) \leq X_j O(C)$. This means that $L_0 \leq X_j O(C) \leq E_j O(C)$ which is a 2-component of C. □

LEMMA 9.4. *Suppose that $\mathcal{L}_2(C) = \{E_1, ..., E_n\}$ for some $n \in \mathbb{N}$, let $i \in \{1, ..., n-1\}$ and let $Y := E_{i+1} \cdots E_n O_{2',2}(C)$. Then C does not possess elementary abelian subgroups A_1 and A_2 of order 4 with the following properties:*

- $A_1 \leq E_1 \cdots E_i$,
- $A_2 \leq C_Y(A_1)$ and
- $A_1 \cap A_2 = 1$.

PROOF. Assume that such subgroups A_1, A_2 exist. Then let $A := A_1 \times A_2$, let $F := E_1 \cdots E_i Y$ and $T \in \mathrm{Syl}_2(O_{2',2}(C))$. Our objective is to show that A_1 and A_2 are core-separated (as in Definition 5.9).

Let $a \in A^\#$ and $H := C_G(a)$. If $L \in \mathcal{L}_2(H)$ is arbitrary and $L_0 := O^\infty(C_L(z))$, then Lemma 9.3 yields that there exists some $j \in \{1, ..., n\}$ such that L_0 is contained in E_j. This implies that

$$[L_0, A_1] \leq O(C) \text{ (if } i < j\text{) or } [L_0, A_2] \leq O(C) \text{ (if } i \geq j\text{)}.$$

As $z \in O_{2',2}(H)$ by Lemma 4.2 (3), it follows that

$$[L_0, A_1] \leq O(C_H(z)) \leq O(H) \text{ or } [L_0, A_2] \leq O(C_H(z)) \leq O(H)$$

with Lemma 2.8. But $z \in Z^*(H)$ also means that $L = L_0 O(L) \leq L_0 O(H)$ and therefore

$$[L, A_1] \leq [L_0 O(H), A_1] \leq O(H) \text{ or } [L, A_2] \leq [L_0 O(H), A_2] \leq O(H).$$

Hence A_1 and A_2 are core-separated and Lemma 5.13 gives a contradiction. □

LEMMA 9.5. *Suppose that $\overline{E}$ is a component of $\overline{C}$ of 2-rank 1 and let $\overline{T} \in \mathrm{Syl}_2(\overline{E})$. Then $\overline{E} \simeq 2A_7$ or there exists some odd number $q \geq 5$ such that $\overline{E} \simeq SL_2(q)$. In particular $\overline{T}$ is quaternion and the unique involution in $\overline{T}$ is $\overline{z}$.*

PROOF. This follows from Lemma 2.22 and from the fact that by Theorem **B** the only involution in $Z(E(\overline{C}))$ is $\overline{z}$. □

LEMMA 9.6. *Suppose that $E \in \mathcal{L}_2(C)$ is such that $z \in E$ and $r_2(E) \geq 2$. Then either $r_2(E) \geq 3$ or there exists a subgroup $W \leq E$ such that $W \simeq C_4 \times C_2$ and $z \in \Phi(W)$.*

PROOF. Let $T \in \mathrm{Syl}_2(E)$. As $z \in Z(E)$, Lemma 4.5 yields that $O^2(C) = C$ and hence $O^2(G) = G$ by Lemma 3.1 (10). Then Lemma 2.17 implies that T is not dihedral or semi-dihedral. It follows from Lemma 2.11 that T has a normal elementary abelian subgroup B of order 4. Set $T_0 := C_T(B)$, $\overline{Z} := Z(\overline{E})$ and $\widetilde{E} := \overline{E}/\overline{Z}$. Moreover let

$$\mathcal{W} := \{W \leq E \mid W \simeq C_4 \times C_2, z \in \Phi(W)\}.$$

Assume that $r_2(E) = 2$ (in particular $z \in B$) and that $\mathcal{W} = \varnothing$. Let $b \in B$ be such that $B = \langle z, b \rangle$. If $\overline{Z} \neq \langle \overline{z} \rangle$, then Theorem **B** yields that Z contains a cyclic subgroup of order 4 and then, as $r_2(E) = 2$, it follows that $\mathcal{W} \neq \varnothing$, contrary to our assumption. Therefore $\overline{Z} = \langle \overline{z} \rangle$. Also $r_2(\widetilde{E}) \geq 2$ because $\widetilde{E}$ is simple. So there exists an element $u \in T$ such that $\widetilde{u}$ is an involution and $\widetilde{u} \notin \widetilde{B}$. Then either u is an involution or $o(u) = 4$, in which case $\langle u, B \rangle \simeq D_8$ because $\mathcal{W} = \varnothing$. Therefore we may suppose that u is an involution. As T is not dihedral or semi-dihedral, Lemma 2.12 implies that $C_T(u)$ has order at least 8. Together with our assumptions that

$r_2(E) = 2$ and $\mathcal{W} = \emptyset$ this yields that $C_T(u)$ contains an element of order 4 that squares to u or to uz. By symmetry we may suppose that $x \in C_T(u)$ is such that $x^2 = u$. As $B \trianglelefteq T$ and $z \in Z(T)$, we see that x centralises b or interchanges b and bz. Therefore u centralises B, but $u \notin B$, and this contradicts the fact that $r(T) = 2$. $\square$

LEMMA 9.7. *Suppose that $E_1, E_2 \in \mathcal{L}_2(C)$ are distinct. Then $r_2(E_1 E_2) \geq 3$.*

PROOF. Assume otherwise. If $r_2(E_1) = r_2(E_2) = 1$, then $\overline{E_1}$ and $\overline{E_2}$ have quaternion Sylow 2-subgroups with central involution $\overline{z}$ by Lemma 9.5. Thus Lemma 2.13 gives a contradiction. We also see that none of the components has 2-rank 3 or more, so we suppose that $r_2(E_1) = 2$. If $z \notin E_1$, then $\overline{E_1}$ is simple by Theorem **B** and hence $\overline{E_1} \cap \overline{E_2} = 1$. Then it follows that $r_2(E_1 E_2) \geq 3$, contrary to our assumption. Hence $z \in E_1 \cap E_2$. Lemma 9.6 yields that E_1 possesses a subgroup W such that $W \simeq C_4 \times C_2$ and $z \in \Phi(W)$. In particular there exists an element $x \in W$ that squares to z. There also exists an element $y \in E_2$ that squares to z (because $r_2(E_2) = 1$ or 2) and we can choose y such that $[W, y] = 1$. Then xy is an involution that is not contained in W and hence $W\langle xy \rangle$ has rank 3, which is a contradiction. $\square$

LEMMA 9.8. *Suppose that $\overline{E_1}$, $\overline{E_2}$ and $\overline{E_3}$ are distinct components of $\overline{C}$. Then these are the only components of $\overline{C}$ and for every $i \in \{1, 2, 3\}$, we have that $\overline{E_i}$ is isomorphic to $2A_7$ or there exists an odd number $q_i \geq 5$ such that $\overline{E_i} \simeq SL_2(q_i)$. Moreover $O_2(\overline{C}) = \langle \overline{z} \rangle \leq E(\overline{C})$.*

PROOF. By Lemma 9.7 we know that $r_2(E_1 E_2) \geq 3$ and therefore there exists an elementary abelian subgroup A of $E_1 E_2$ of order 4 that does not contain z. If $r_2(E_3) \geq 2$, then $C_{E_3}(A)$ contains an elementary abelian subgroup B of order 4 that intersects A trivially, contrary to Lemma 9.4. Therefore $r_2(E_3) = 1$. By symmetry we deduce, for all $E \in \mathcal{L}_2(C)$, that $r_2(E) = 1$.

Assume that there exists a 2-component $L \in \mathcal{L}_2(C) \setminus \{E_1, E_2, E_3\}$ or that $O_{2',2}(C)$ has a Sylow 2-subgroup of order at least 4. Lemma 9.4 implies that these cases cannot occur both at once, because $r_2(E_1 E_2 E_3) \geq 3$. Therefore we let $T \in \mathrm{Syl}_2(L)$ or $T \in \mathrm{Syl}_2(O_{2',2}(C))$, respectively. We recall that $r_2(E_3) = 1$ and therefore $z \in E_3$ by Theorem **B**. This theorem also implies that z is the only square in T and in E_3, therefore TE_3 contains diagonal involutions and in particular $r_2(TE_3) \geq 2$. With A as in the previous paragraph, it follows that $C_{TE_3}(A)$ contains an elementary abelian subgroup of order 4 that intersects A trivially. Again we have a contradiction to Lemma 9.4. Together with Lemma 9.5 this completes the proof. $\square$

PROOF OF THEOREM **C**.

Recall that in Theorem **C** we suppose that G is a minimal counter-example to the Z*-Theorem. Therefore Hypothesis 4.1 holds. Theorem 9.1 yields that $\overline{C}$ has at least one component and by Lemma 9.8 there are at most three components in $\overline{C}$. If $\overline{C}$ has precisely three components $\overline{E_1}$, $\overline{E_2}$ and $\overline{E_3}$, then with Lemma 9.8 we have for all $i \in \{1, 2, 3\}$ that $\overline{E_i}$ is isomorphic to $2A_7$ or that there exists an odd number $q_i \geq 5$ such that $\overline{E_i}$ is isomorphic to $SL_2(q_i)$.

This completes the proof of Theorem **C**. $\square$

For the Soluble Z*-Theorem we do not need the full force of Theorem **C**. It is sufficient that, if Hypothesis 4.1 holds, then $\mathcal{L}_2(C) \neq \varnothing$.

PROOF OF THE SOLUBLE Z*-THEOREM.

Assume that G is a minimal counter-example to the Soluble Z*-Theorem. Let $z \in G$ be an isolated involution such that $C := C_G(z)$ is soluble and assume that $z \notin Z^*(G)$. If $z \in H < G$, then the choice of G as a minimal counter-example yields that $H = C_H(z)O(H)$ and thus H is soluble. Let $t \in G$ be an arbitrary involution. Lemma 3.1 (2) and Sylow's Theorem imply that $C_G(t)$ contains a conjugate of z. Thus $C_G(t)$ is soluble by the previous paragraph. From the minimality of G and the fact that every involution centraliser is soluble, it follows that the Z*-Theorem holds in every proper subgroup and every proper section of G. This means that Hypothesis 4.1 is satisfied. In particular, Theorem 9.1 is applicable and yields that $C/O(C)$ has at least one component. This is impossible because C is soluble. □

CHAPTER 10

Unbalanced Components

In this section we prepare the proof of Theorem **D** and here it becomes necessary to invoke knowledge about the simple groups involved in $E(\overline{C})$ and their automorphism groups. Whenever we use specific information about the components in $\overline{C}$ (usually the 2-structure, involutions centralisers or automorphisms), then these details are from [**GLS98**], more precisely from Tables 3.3.1, 4.5.1 and 4.5.2, Theorems 4.10.5 and 5.2.1, Proposition 5.2.10 and Tables 5.3a, 5.3g and 5.6.1, from the ATLAS [**CCN103**] or from the corresponding sections in [**Wil09**]. In some places we make this more precise. Here comes our new main hypothesis:

HYPOTHESIS 10.1. *In addition to Hypothesis* 4.1, *the components of $\overline{C}$ are supposed to be known quasi-simple groups.*

The following remark captures a general fact about Lie type groups that is used in connection with balance arguments or in determining types of components. Then we introduce the notion of an unbalanced component and state the consequences of results from Chapter 4 in [**Gor82**] for our situation.

REMARK 10.2. Suppose that E is a quasi-simple group of Lie type in odd characteristic. If $t \in E$ is an involution and $x \in \operatorname{Aut}(E)$ is a non-trivial field automorphism of E of odd order, then $C_E(t) \not\leq C_E(x)$.

(The involution is contained in a torus, so this torus lies in $C_E(t)$, but not in $C_E(x)$.)

DEFINITION 10.3. A component $\overline{E}$ of $\overline{C}$ is called an **unbalanced component** if and only if it is of type A_n (with $n \in \mathbb{N}$ and $n \equiv 3$ modulo 4) or of type $PSL_2(q)$ (with an odd number $q \geq 5$).

LEMMA 10.4. *Suppose that Hypothesis* 10.1 *holds and that $r_2(G) \geq 4$. Then $\overline{C}$ possesses an unbalanced component. Moreover every component of $\overline{C}$ of 2-rank at least 4 is unbalanced of type A_n (with $n \equiv 3$ modulo 4 and $n \geq 11$).*

PROOF. We refer to Section 4.4 in [**Gor82**]. By Hypothesis 10.1, the simple groups involved in $E(\overline{C})$ are known. Moreover Lemma 5.7 yields that C does not possess any 2-balanced subgroups. Therefore by Proposition 4.64 in [**Gor82**], some component of $\overline{C}$ is not locally 2-balanced (as defined there). Then $\overline{C}$ possesses an unbalanced component by Theorem 4.61 in [**Gor82**].

If $E \in \mathcal{L}_2(C)$ has 2-rank at least 4, then Lemma 5.7 yields that E does not have any 2-balanced subgroups and hence $\overline{E}$ is unbalanced by Proposition 4.64 and Theorem 4.61 in [**Gor82**]. Then there exists some $n \in \mathbb{N}$ such that $n \equiv 3$ modulo 4 and $n \geq 11$ by Theorem 2.16. □

LEMMA 10.5. *Suppose that Hypothesis 10.1 holds and let $n \in \mathbb{N}$ be such that $n \geq 10$. Then $\overline{C}$ does not possess a component of type A_n.*

PROOF. Assume that $n \in \mathbb{N}$ and $\overline{E} \in \mathcal{L}_2(C)$ are such that $n \geq 10$ and $\overline{E}$ is a component of $\overline{C}$ of type A_n. If $\overline{E}$ is not normal in $\overline{C}$, then we find elementary abelian subgroups of order 16 in $\overline{E}$ and in some conjugate of $\overline{E}$, respectively, by Theorem 2.16. The intersection of these subgroups lies in $\langle \overline{z} \rangle$ (with Theorem **B**). Hence Lemma 9.4 yields a contradiction and it follows that $\overline{E}$ is normal in $\overline{C}$.

(1) $\overline{C}$ induces inner automorphisms on $\overline{E}$.

PROOF. Let $S \in \mathrm{Syl}_2(C)$ and $T := S \cap O^2(G)$. Then Lemma 4.6 yields that either $O^2(\overline{C}) = \overline{C}$ and hence $S = T$ or $|\overline{C} : O^2(\overline{C})| = 2$ and hence $S = T \times \langle z \rangle$, by Lemma 4.5. As E is normal in C and z centralises E, we deduce that $\overline{C}/C_{\overline{C}}(\overline{E})$ is isomorphic to a subgroup of $O^2(\mathrm{Aut}(\overline{E}))$ and then Theorem 2.16 implies that $\overline{C}$ induces inner automorphisms on $\overline{E}$. □

(2) One of the following holds:
  - $\overline{E}$ is the unique component of $\overline{C}$ or
  - $\overline{C}$ possesses precisely two components $\overline{E}$ and $\overline{L}$ and then $r_2(\overline{L}) = 1$ and $F^*(\overline{C}) = E(\overline{C})$.

PROOF. Suppose that there exists another component $\overline{L}$. As $n \geq 10$, Theorem 2.16 yields that there exists an elementary abelian subgroup $A_1 \leq E$ of order 4 that does not contain z. Thus if $r_2(L) > 1$, then there exists an elementary abelian subgroup $A_2 \leq C_L(A_1)$ of order 4 such that $A_1 \cap A_2 = 1$, contrary to Lemma 9.4. Hence $\overline{L}$ is of 2-rank 1 as stated.

Assume that $O_2(\overline{C}) > \langle \overline{z} \rangle$ and let $z \in T \in \mathrm{Syl}_2(O_{2',2}(C))$. Then $\overline{T} = O_2(\overline{C})$ and T possesses elements of order 4 (because z is the only involution in T by Theorem **B**). As all elements of order 4 in L and T square to z, we have diagonal involutions in TL and therefore $C_{TL}(A_1)$ contains an elementary abelian subgroup of order 4. Again this contradicts Lemma 9.4 and we conclude that $O_2(\overline{C}) = \langle \overline{z} \rangle$. Now we have that either $\overline{E}$ is the unique component of $\overline{C}$ or $F^*(\overline{C}) = \overline{EL}$ in which case $\overline{E}$ and $\overline{L}$ are normal in $\overline{C}$ and $C_{\overline{C}}(F^*(\overline{C})) = \langle \overline{z} \rangle$. □

Set $\tilde{C} := \overline{C}/\langle \overline{z} \rangle$. We choose involutions a_1, a_2, a_3 of E such that, for all $i \in \{1, 2, 3\}$, the following holds:
$$O(C_{\tilde{E}}(\tilde{a}_i)) = 1$$
or
$$O(C_{\tilde{E}}(\tilde{a}_i)) = O(C_{\tilde{E}}(\tilde{A})).$$

Let $a_1 \in E$ be such that $\tilde{a}_1 \in \tilde{E}$ corresponds to the element $(12)(34)(56)(78)$ in A_n. Let $a_2 \in E$ be such that $\tilde{a}_2 \in \tilde{E}$ corresponds to the element $(13)(24)(57)(68)$ in A_n and let $a_3 := a_1 a_2$.

By Theorem 33.15 in [**Asc00**], the elements $\overline{a_1}$, $\overline{a_2}$ and $\overline{a_3}$ are commuting involutions in $\overline{E}$. As $C_{\overline{E}}(\overline{a_i}) = \overline{C_E(a_i)}$, we can choose a_1, a_2 and a_3 to be commuting involutions in E. It follows that $A := \langle a_1, a_2, z \rangle$ is an elementary abelian subgroup of order 8 of C. Unless $n = 11$, we have for all $i \in \{1, 2, 3\}$ that $O(C_{\tilde{E}}(\tilde{a}_i)) = 1$ as required in the first case above. When $n = 11$, the second case holds because then, for all involutions $a \in A\setminus\{1,z\}$, the subgroup $O(C_{\tilde{E}}(\tilde{a})) = O(C_{\tilde{E}}(\tilde{A}))$ is cyclic of order 3.

(3) A is weakly balanced.

PROOF. Let $i \in \{1,2,3\}$. By Lemma 5.5 (1) it suffices to prove that $\alpha(a_i) \leq O(C)O(C_C(A))$. From (1) we know that $\overline{\alpha(a_i)}$ induces inner automorphisms on $\overline{E}$, and Lemma 5.15 yields that $\overline{\alpha(a_i)}$ centralises $O_2(\overline{C})$ and, if it exists, the second component of $\overline{C}$. So we deduce that $\overline{\alpha(a_i)}$ induces inner automorphisms on $F^*(\overline{C})$ and is therefore contained in $F^*(\overline{C})$. It follows that $\overline{\alpha(a_i)} = O(C_{\overline{E}}(\overline{a_i}))$ and then our choice of A implies that

$$\overline{\alpha(a_i)} = 1 \text{ or } \overline{\alpha(a_i)} = O(C_{\overline{C}}(\overline{A})).$$

This forces $\alpha(a_i) \leq O(C)O(C_C(A))$ for all $i \in \{1,2,3\}$, in both cases, as required. We conclude that A is weakly balanced. □

Now Lemma 5.8 yields a contradiction. □

COROLLARY 10.6. *Suppose that Hypothesis* 10.1 *holds and that* $r_2(G) \geq 4$. *Then $\overline{C}$ possesses a component of type* A_7 *or of type* $PSL_2(q)$ *(with an odd number* $q \geq 5$*)*.

PROOF. By hypothesis and Theorem 10.4 we know that $\overline{C}$ possesses an unbalanced component. But in the case of type A_n, Lemma 10.5 forces n to be at most 9. As A_3 is soluble, the only possible component type left is A_7. □

CHAPTER 11

The 2-Rank of G

In this section we prove Theorem **D**, one of the most important results towards understanding the structure of $F^*(\overline{C})$. But even to establish Theorem **D** we already need some knowledge about the components of $\overline{C}$. In order to make the arguments more clear, we begin by excluding a few particular configurations, thus dealing with some of the technical details separately.

LEMMA 11.1. *Suppose that Hypothesis* 10.1 *holds and that* $E \in \mathcal{L}_2(C)$. *Suppose further that* $\overline{E} \simeq A_7$ *or that there exists an odd number* $q \geq 5$ *such that* $\overline{E} \simeq PSL_2(q)$. *Then* $F^*(\overline{C}) \neq \langle \overline{z} \rangle \overline{E}$.

PROOF. Assume otherwise, which means that $F^*(\overline{C}) = \langle \overline{z} \rangle \overline{E}$ with the type of $\overline{E}$ as stated. Then $\overline{E}$ is simple and therefore $\overline{z} \notin \overline{E}$, so we see that actually $F^*(\overline{C}) = \langle \overline{z} \rangle \times \overline{E}$. In particular $O^2(C) \neq C$ and therefore $O^2(G) \neq G$ by Lemma 3.1 (10). Let $S_0 \in \mathrm{Syl}_2(E)$. Then z centralises S_0, but $z \notin S_0$ and thus $S_1 := \langle z \rangle \times S_0$ is a Sylow 2-subgroup of $O_{2',F^*}(C)$. Let $S_1 \leq S \in \mathrm{Syl}_2(G)$. Then $S \leq C$ by Lemma 3.1 (2) and $S_1 = S \cap O_{2',F^*}(C) \trianglelefteq S$, in particular $\overline{S_1} \trianglelefteq \overline{S}$. Every element of $\overline{S}$ outside $\overline{S_1}$ centralises $\overline{z}$, but is not contained in $F^*(\overline{C})$ and therefore induces an outer automorphism on $\overline{E}$.

Now we assume that $S \not\leq O_{2',F^*}(C)$ which means that $S_1 < S$. Then we recall that, by the first paragraph, the elements from $\overline{S} \backslash \overline{S_1}$ induce outer automorphism on $\overline{E}$. The outer automorphism group of $\overline{E}$ is 2-nilpotent (because of the type of $\overline{E}$), and we also know that $\overline{z} \notin O^2(\overline{C})$, therefore $|\overline{C} : O^2(\overline{C})| \geq 4$. But this contradicts Lemma 4.6.

We conclude that $S_1 = S \in \mathrm{Syl}_2(G)$ and this means that S is a direct product of $\langle z \rangle$ with the dihedral group S_0 (that could be a fours group). As S_0 is dihedral, the Gorenstein-Walter Theorem 2.21 yields that $F^*(G)$ is isomorphic to A_7 or that there exists an odd number $q' \geq 5$ such that $F^*(G) \simeq PSL_2(q')$. As $\mathcal{S}_7$ does not have any isolated involutions, we cannot have that $G \simeq \mathcal{S}_7$ and therefore $F^*(G)$ is not isomorphic to A_7. We are left with the case that $F^*(G) \simeq PSL_2(q')$. We know that $z \notin F^*(G)$ and hence z induces an outer automorphism on $F^*(G)$. This must be a field automorphism because z centralises a Sylow 2-subgroup of $F^*(G)$. Let $C_0 := C_{F^*(G)}(z)$. Then there exists a prime power q_0 dividing q' such that $C_0 \simeq PGL_2(q_0)$. In particular $O^2(C_0) \neq C_0$, contrary to Lemma 4.6. □

LEMMA 11.2. *Suppose that Hypothesis* 10.1 *holds and that* $E \in \mathcal{L}_2(C)$. *Then* $\overline{E}$ *cannot be isomorphic to any of the following:*
 – $Sp_6(q)$ *with an odd number* q;
 – $2J_2$; *or*
 – M_{11}.

PROOF. Assume that $\overline{E}$ is isomorphic to one of the groups listed and note that, in the first two cases, this implies that $\overline{E}$ is not simple. Thus $\overline{E}$ contains a central involution which by Theorem **B** can only be $\overline{z}$. We also observe that E has 2-rank at least 3 in the first case because $Sp_6(q)$ contains a subgroup isomorphic to $SL_2(q) \times SL_2(q) \times SL_2(q)$. In the second case we have that $r_2(E) = 3$ because $r_2(2J_2) = 3$ and in the last case we have that $r_2(E) = 2$, see for example Table 5.6.1 in [**GLS98**]. Let B be an elementary abelian subgroup of order 4 of E that does not contain z (as is possible in all cases), so that $A := B\langle z\rangle$ is elementary abelian of order 8. Let $b \in B^\#$. With Tables 4.5.2, 5.3g and 5.3a in [**GLS98**], we see that $O(C_{\overline{E}}(\overline{b})) = 1$. Hence if $\alpha(b) \not\leq O(C)$, then Lemma 5.15 implies that $\overline{\alpha(b)}$ induces an outer automorphism of $\overline{E}$ of odd order. This is impossible in the J_2-case and in the M_{11}-case (see Tables 5.3g and 5.3a in [**GLS98**]). In the remaining case $\overline{\alpha(b)}$ induces a field automorphism. But we also know that

$$[\overline{\alpha(b)}, C_{\overline{E}}(\overline{b})] \leq O(C_{\overline{C}}(\overline{b})) \cap \overline{E} \leq O(C_{\overline{E}}(\overline{b})) = 1.$$

Then $C_{\overline{E}}(\overline{b}) \leq C_{\overline{E}}(\overline{\alpha(b)})$ and this is impossible by Remark 10.2.

It follows that $\alpha(b) \leq O(C)$ and hence Lemma 5.5 (1) yields, for all $a \in A^\#$, that $\alpha(a) \leq O(C)$. Thus A is balanced, which is a contradiction to Lemma 5.8. □

LEMMA 11.3. *Suppose that Hypothesis* 10.1 *holds and that* $E \in \mathcal{L}_2(C)$. *Then* $\overline{E}/Z(\overline{E})$ *is not isomorphic to* $PSL_3(4)$.

PROOF. Assume otherwise. We refer to the ATLAS ([**CCN103**]) for information about $PSL_3(4)$. If $Z(\overline{E}) = 1$, then $\overline{E} \simeq PSL_3(4)$ and this is impossible by Lemma 10.4. Hence $Z(\overline{E})$ is a non-trivial 2-group with unique involution $\overline{z}$, by Theorem **B**. The Schur multiplier leaves the possibilities $2PSL_3(4)$ and $4PSL_3(4)$. As $2PSL_3(4)$ contains an elementary abelian group of order 16, it is excluded by Lemma 10.4 as well. Hence $\overline{E} \simeq 4PSL_3(4)$ and in particular $r_2(E) = 3$. Let B be an elementary abelian subgroup of order 4 of E that does not contain z and let $A := B\langle z\rangle$. Then A is elementary abelian of order 8. Let $b \in B^\#$. In $PSL_3(4)$ the centralisers of involutions are 2-groups, and so it follows that $O(C_{\overline{E}}(\overline{b})) = 1$. Hence if $\alpha(b) \not\leq O(C)$, then $\overline{\alpha(b)}$ induces an outer automorphism of $\overline{E}$ of odd order, by Lemma 5.15. This must be an automorphism of order 3. But since $Z(\overline{E})$ has only order 4, this is not possible. We conclude that $\alpha(b) \leq O(C)$ for all $b \in B^\#$ (and hence for all $b \in A^\#$), so the subgroup A is balanced. This contradicts Lemma 5.8. □

LEMMA 11.4. *Suppose that Hypothesis* 10.1 *holds. Then* $\overline{C}$ *does not have a simple component of 2-rank 3.*

PROOF. Assume that $E \in \mathcal{L}_2(C)$ is of 2-rank 3 and such that $\overline{E}$ is simple. We know from Hypothesis 10.1 that $\overline{E}$ is isomorphic to an alternating group, to a Lie type group or to a sporadic group.

$\overline{E}$ is not isomorphic to an alternating group by Theorem 2.16 (2). If $\overline{E}$ is isomorphic to a group of Lie type in characteristic 2, then Table 3.3.1 in [**GLS98**] implies that $\overline{E} \simeq L_2(8)$, $Sz(8)$ or $U_3(8)$. If $\overline{E}$ is isomorphic to a group of Lie type in odd characteristic, then Theorem 4.10.5 in [**GLS98**] yields that there exists an odd number q such that $\overline{E}$ is isomorphic to $G_2(q)$, to $^2G_2(q)$ or to $^3D_4(q)$, because we excluded the case $Sp_6(q)$ in Lemma 11.2. If $\overline{E}$ is isomorphic to a sporadic group, then Table 5.6.1 in [**GLS98**] leaves the possibilities that $\overline{E}$ is isomorphic to M_{12}, to J_1 or to $O'N$, because we excluded the possibilities $2J_2$ and M_{11} in Lemma 11.2.

Now we let $a \in E$ be an involution and we inspect pages 6, 28 and 66 in the ATLAS ([**CCN103**]) for the cases in characteristic 2 and Tables 4.5.1, 5.3b, 5.3f and 5.3s in [**GLS98**] for the remaining cases. Then we see that $O(C_{\overline{E}}(\overline{a})) = 1$. If $x \in C$ is such that $1 \neq \overline{x} \in \overline{\alpha(a)} = O(C_{\overline{C}}(\overline{a}))$, then Lemma 5.15 forces $\overline{x}$ to induce an outer automorphism of odd order on $\overline{E}$. This leaves only the Lie type cases, again by inspection of Tables 5.3b, 5.3f and 5.3s. But, as before, we have that

$$[\overline{x}, C_{\overline{E}}(\overline{a})] \leq O(C_{\overline{C}}(\overline{a})) \cap \overline{E} \leq O(C_{\overline{E}}(\overline{a})) = 1$$

and therefore $C_{\overline{E}}(\overline{a}) \leq C_{\overline{E}}(\overline{x})$. In the Lie type cases in odd characteristic, this contradicts Remark 10.2. In the three cases with characteristic 2, we also see (for example in the ATLAS) that $C_{\overline{E}}(\overline{a}) \not\leq C_{\overline{E}}(\overline{x})$ and so we have a contradiction there as well.

It follows that $O(C_{\overline{C}}(\overline{a})) = 1$. As $a \in E$ was an arbitrary involution and $r_2(E) = 3$, we just proved that C contains a 2-subgroup A of rank 4 containing z that is balanced. This contradicts Lemma 5.8. □

DEFINITION 11.5. Let q be a power of an odd prime.
List (a): $2A_7$, $SL_2(q)$, $SL_4(q)$, $Sp_4(q)$, $SU_4(q)$.
List (b): A_7, $PSL_2(q)$, $PSL_3(q)$, $PSU_3(q)$, $PSU_3(4)$.

LEMMA 11.6. *Suppose that Hypothesis 10.1 holds. Let $E \in \mathcal{L}_2(C)$ and suppose that $r_2(E) \leq 3$. Then $\overline{E}$ is isomorphic to a group from List (a) or (b).*

PROOF. By Hypothesis 10.1, we must consider the following cases: $\overline{E}$ is of alternating type, of Lie type or sporadic.

We also point out that, if $\overline{E}$ is simple, then $\overline{E}$ has 2-rank 2 by Lemmas 9.5 and 11.4. Suppose that $\overline{E}$ is of alternating type. Then our hypothesis about the 2-rank and Theorem 2.16 imply that $\overline{E}$ can only be of types A_5, A_6 or A_7. These are on the lists (recall that $A_5 \simeq PSL_2(5)$ and $A_6 \simeq PSL_2(9)$).

Next suppose that $\overline{E}$ is of Lie type. If $\overline{E}$ is defined in characteristic 2, then we first inspect the ATLAS ([**CCN103**]) and Table 3.3.1 in [**GLS98**]. This leads to the possibilities that $\overline{E}$ is of type $PSL_2(4) \simeq PSL_2(5)$, of type $PSL_3(2) \simeq PSL_2(7)$, of type $Sp_4(2)' \simeq PSL_2(9)$, or of types $PSU_3(4)$ or $PSU_3(3)$. These groups are on the lists above. We need to look at exceptional Schur multipliers (for example Table 6.1.3 in [**GLS98**]) as well, bearing in mind Theorem **B**. We already considered the types $PSL_2(4)$ and $PSL_3(2)$, Lemma 11.3 excludes type $PSL_3(4)$, the type $PSL_4(2) \simeq A_8$ does not occur, and $PSU_4(2) \simeq Sp_4(3)$ is on List (a). The other groups with exceptional Schur multipliers do not appear because otherwise the 2-rank of E is too large.

If $\overline{E}$ is defined in odd characteristic, then Theorem 4.10.5 in [**GLS98**] and Lemma 11.2 only leave possibilities from the lists. (Groups like $SL_3(q)$ do not occur because $Z(E(\overline{C}))$ is a 2-group, and exceptional Schur multipliers do not play any role here.)

Finally $\overline{E}$ cannot be of sporadic type by Table 5.6.1 in [**GLS98**] and Lemma 11.2. □

PROOF OF THEOREM **D**.
Recall that in Theorem **D**, we suppose that Hypothesis 10.1 holds. Assume that the theorem is false. Then $r_2(G) \geq 4$ and therefore, by Corollary 10.6, there exists a component in $\overline{C}$ that is of type A_7 or, for a suitable odd number $q \geq 5$, of type $PSL_2(q)$. We also know from Theorem **B** that $O_2(\overline{C})$ is cyclic or quaternion. Let $E_1 \in \mathcal{L}_2(C)$ be such that $\overline{E_1}$ is of type A_7 or $PSL_2(q)$. If $\overline{E_1}$ is not simple, then the Sylow 2-subgroups of $\overline{E_1}$ are quaternion whence precisely one of the following holds:
- $O_2(\overline{C}) = \langle \overline{z} \rangle = Z(\overline{E_1})$ or
- $r_2(\overline{E_1}O_2(\overline{C})) \geq 2$ (because there exist diagonal involutions).

In the following it happens several times that we deduce from the structure of $F^*(\overline{C})$ that a Sylow 2-subgroup $\overline{S}$ of $\overline{C}$ induces inner automorphisms on (and hence lies in) $F^*(\overline{C})$. More specifically, if the outer automorphism group of $F^*(\overline{C})$ is 2-nilpotent (as is the case if we have at most two components), then $\overline{S}$ induces inner automorphisms on $F^*(\overline{C})$ by Lemma 4.6, because $\overline{z}$ centralises $F^*(\overline{C})$.

For the remainder of the proof, we fix $\overline{S} \in \mathrm{Syl}_2(\overline{C})$ and we refer to the above argument by saying that the structure of $F^*(\overline{C})$ forces $\overline{S}$ to be contained in $F^*(\overline{C})$.

Case 1: $\overline{E_1}$ is the only component of $\overline{C}$.

First assume that $\overline{E_1}$ is simple. Then $\overline{E_1}$ has dihedral Sylow 2-subgroups, because of its type, and therefore $r_2(\overline{E_1}) = 2$. It follows that $r_2(F^*(\overline{C})) = r_2(\overline{E_1}O_2(\overline{C})) = 3$. The structure of $F^*(\overline{C})$ forces $\overline{S}$ to be contained in $F^*(\overline{C})$ and therefore $\overline{S}$ is of rank 4, but this is a contradiction.

It follows that $\overline{E_1}$ is not simple, therefore $Z(\overline{E_1})$ contains an involution and by Theorem **B** this involution must be $\overline{z}$. Again the structure of $F^*(\overline{C})$ forces $\overline{S}$ to be contained in $F^*(\overline{C})$. But then $\overline{S}$ is a central (and non-direct) product of a quaternion group with either a cyclic group or another quaternion group and therefore $\overline{S}$ has rank 2 or 3. This is again a contradiction.

Case 2: $\overline{C}$ has exactly two components.

Let $E_2 \in \mathcal{L}_2(C)$ be such that $\overline{E_2}$ is the second component of $\overline{C}$. If $r_2(\overline{E_2}) \geq 4$, then Lemma 10.4 forces $\overline{E_2}$ to be unbalanced of type A_n (with $n \equiv 3$ modulo 4 and $n \geq 11$). But this is excluded by Lemma 10.5. Hence $r_2(\overline{E_2}) \leq 3$.

Assume first that $\overline{E_1}$ is simple. Then $\overline{E_1}$ has 2-rank 2 because of its type. Moreover we have that $\overline{E_1} \cap \overline{E_2}O_2(\overline{C}) = 1$. Therefore $\overline{E_2}O_2(\overline{C})$ is of 2-rank at most 1 because otherwise Lemma 9.4 yields a contradiction. We deduce that $r_2(F^*(\overline{C})) = 3$ and that $\overline{E_1}$ and $\overline{E_2}$ are normal in $\overline{C}$. Therefore the structure of $F^*(\overline{C})$ forces $\overline{S}$ to be contained in $F^*(\overline{C})$, which is a contradiction. Thus $\overline{E_1}$ is not simple. If $\overline{E_2}$ is simple, then Lemma 11.4 together with the fact that $r_2(\overline{E_2}) \leq 3$ implies that $r_2(\overline{E_2}) = 2$ and that, therefore, $\overline{E_2}$ is isomorphic to a group from List (b) in Definition 11.5. Moreover $\overline{E_1}$ and $\overline{E_2}$ are normal in $\overline{C}$, so again the structure of $F^*(\overline{C})$ forces $F^*(\overline{C})$ to contain a Sylow 2-subgroup of $\overline{C}$. It follows that $r_2(O_{2',F^*}(C)) = r_2(G) \geq 4$, and this is only possible if $r_2(\overline{E_1}O_2(\overline{C})) \geq 2$. But $\overline{E_1}O_2(\overline{C}) \cap \overline{E_2} = 1$ and therefore this contradicts Lemma 9.4. We conclude that $\overline{E_1}$ and $\overline{E_2}$ are both non-simple and that $\overline{S} \leq F^*(\overline{C})$. This is impossible if $\overline{E_2}$ is isomorphic to $2A_7$ or $SL_2(q)$, because then the 2-rank of $F^*(\overline{C})$ is only 3 (recall Lemma 2.13 and the fact that $O_2(\overline{C})$ is cyclic or quaternion). In the remaining cases from List (a) there exists an odd number q such that $\overline{E_2} \simeq SL_4(q'), Sp_4(q')$ or $SU_4(q')$. Therefore $\overline{E}$ has 2-rank at least 2 and Lemma 9.4 forces $r_2(\overline{E_1}O_2(\overline{C})) \leq 2$. This means that $O_2(\overline{C})$ is cyclic, with Lemma 2.13.

Let U be the 4-dimensional module over a field of order q' defining $\overline{E_2}$ and let U_1, U_2 be 2-dimensional subspaces of U such that U is the direct orthogonal sum of U_1 and U_2. Let $a \in S$ be an involution such that $\overline{a} \in \overline{E_2}$ and such that U_1 and U_2 are the eigenspaces of $\overline{a}$. Then $C_{\overline{E_2}}(\overline{a})$ has subgroups $\overline{L_1}$ and $\overline{L_2}$ such that $\overline{L_1} \times \overline{L_2} \trianglelefteq C_{\overline{E_2}}(\overline{a})$ and such that, for all $i \in \{1,2\}$, the subgroup $\overline{L_i}$ acts faithfully as $SL_2(q')$ on U_i and centralises U_{3-i}. As $O(C_{\overline{E_2}}(\overline{a}))$ acts by scalar multiplication on U_1 and U_2, it centralises $\overline{L_1} \times \overline{L_2}$. Let $L \leq C$ be such that $\overline{L} \simeq SL_2(q')$ and such that $\overline{L}$ is diagonally embedded in $\overline{L_1} \times \overline{L_2}$. Let $T_1 := S \cap E_1$ and $T_2 := S \cap L$. Then T_1 and T_2 are quaternion with common central involution z, so Lemma 2.13 yields that T_1T_2 contains an elementary abelian subgroup B of order 8. As a centralises B, but is not contained in it, it follows that $A := B\langle a \rangle$ is elementary abelian of order 16. The subgroup $O(C_{\overline{E_2}}(\overline{a}))$ of $\overline{E_2}$ centralises $\overline{E_1}$ and $\overline{L_1} \times \overline{L_2}$ and therefore $[O(C_{\overline{E_2}}(\overline{a})), \overline{A}] = 1$.

With the notation from Definition 5.1, we prove, for all subgroups V of order 4 of A, that
$$\Delta_V \leq O(C)O(C_C(A)). \qquad (*)$$

PROOF. Let $V \leq A$ be of order 4. If $z \in V$, then $\Delta_V \leq O(C)$. Also, if $a \in V$, then $\Delta_V \leq O(C_G(a)) \cap C \leq O(C_C(a))$. As $\overline{E_1} \leq E(C_{\overline{C}}(\overline{a}))$, it follows in this case that $\overline{\Delta_V}$ centralises $\overline{F_1}$ and $C_{\overline{E_2}}(\overline{a})$. Thus
$$\overline{\Delta_V} \leq C_{\overline{C}}(\overline{A}) \cap O(C_{\overline{C}}(\overline{a})) \leq O(C_{\overline{C}}(\overline{A}))$$
which implies that $\Delta_V \leq O(C)O(C_C(A))$. A similar argument yields the statement if $az \in V$, so we may suppose that $V \cap \langle a, z \rangle = 1$. As $V\langle z \rangle$ centralises Δ_V, we may also suppose that $[\Delta_V, a] \not\leq O(C)$.

For all $i \in \{1, 2\}$ let $v_i, w_i \in E_i$ be elements of order 4 such that $v := v_1 v_2$, $w := w_1 w_2$ and vw are the involutions in V. As Δ_V centralises V and normalises E_1 and E_2, it follows that $\overline{\Delta_V}$ centralises $\langle \overline{v_1}, \overline{w_1} \rangle$ and $\langle \overline{v_2}, \overline{w_2} \rangle$. Now we set $\widetilde{C} := \overline{C}/Z(\overline{E_2})$, and we note first that $C_{\widetilde{E_1}}(\widetilde{V}) \leq C_{\widetilde{C}}(\widetilde{A})$ because $\widetilde{A} = \widetilde{V}\langle \widetilde{a}, \widetilde{z} \rangle$ and $\widetilde{E_1}$ centralises $\widetilde{a}$. Thus
$$[\widetilde{\Delta_V}, C_{\widetilde{E_1}}(\widetilde{V})] \leq C_{\widetilde{C}}(\widetilde{A}).$$

As $\widetilde{E_2}$ is simple and of 2-rank at least 4 (see for example Theorem 4.10.5 in [**GLS98**]), Theorem 4.61 in [**Gor82**] implies that $C_{\widetilde{E_2}}(\widetilde{V}) = 1$ and therefore
$$[\widetilde{\Delta_V}, C_{\widetilde{E_1 E_2}}(\widetilde{V})] \leq C_{\widetilde{C}}(\widetilde{A}).$$

Thus $[\widetilde{\Delta_V}, \widetilde{A}, \widetilde{A}] \leq [\widetilde{\Delta_V}, C_{\widetilde{E_1 E_2}}(\widetilde{V}), \widetilde{A}] = 1$ and with Lemma 2.1 (2) it follows first that $\overline{\Delta_V} \leq C_{\overline{C}}(\overline{A})$ and then that $\Delta_V \leq O(C)O(C_C(A))$. Thus the proof of $(*)$ is finished.

□

Now we show that Θ (as in Definition 5.1) defines a soluble A-signalizer functor. Let $u \in A^\#$. Then $\Theta(u) = [C_G(u), z]C_{O(C)}(u)O(C_C(A))$ is an A-invariant $2'$-subgroup of $C_G(u)$ and so we are left with the balance condition. Let $w \in A^\#$. With Lemma 2.1 (3), applied twice, we have that

$$\Theta(u) \cap C_G(w)$$
$$= C_{[C_G(u),z]}(w) \cdot (C_{O(C)}(u) \cap C_G(w)) \cdot (O(C_C(A)) \cap C_G(w))$$
$$\leq C_{[C_G(u),z]}(w)C_{O(C)}(w)O(C_C(A)).$$

The second and third subgroup are already contained in $\Theta(w)$, so let $X := C_{[C_G(u),z]}(w)$. As z acts coprimely on X, Lemma 2.1 (2) gives that $X = C_X(z)[X, z]$. We immediately have that $[X, z] \leq [C_G(w), z] \leq \Theta(w)$, so now we look at $C_X(z)$.

Observing that $C_X(z) = C_{[C_G(u),z]}(\langle z\rangle\langle w\rangle)$, we apply Theorem 2.7 to $[C_G(u), z]$, $\langle z \rangle$ and $\langle w \rangle$ in the roles of X, A_0 and B. Then we obtain that

$$C_X(z) = \langle [C_{[C_G(u),z]}(V), z] \cap C_X(z) \mid V \leq A, |V| = 4 \rangle.$$

Let $V \leq A$ be a subgroup of order 4. Then with $(*)$ first and another application of Lemma 2.1 (3) afterwards it follows that

$$[C_{[C_G(u),z]}(V), z] \cap C_X(z) \leq \bigcap_{v \in V^\#} [C_G(v), z] \cap C_X(z) \leq \Delta_V \cap X$$

$$\leq O(C)O(C_C(A)) \cap C_G(w) \leq C_{O(C)}(w)O(C_C(A)) \leq \Theta(w).$$

Thus the balance condition is established and we arrive at a contradiction to Lemma 5.6.

Case 3: $\overline{C}$ has three components.

This case leads to the situation in Lemma 9.8 and in particular $z \in \Phi(S)$. Thus Lemma 4.5 yields that $O^2(C) = C$ and hence $O^2(\overline{C}) = \overline{C}$. Our assumption that $r_2(G) \geq 4$ and the fact that $r_2(F^*(\overline{C})) = 3$ imply that $F^*(\overline{C})$ does not contain a Sylow 2-subgroup of $\overline{C}$. But since $O^2(\overline{C}) = \overline{C}$, this leads to the following configuration (see also Lemma 13.2 in Chapter 13): C is transitive on $\mathcal{L}_2(C)$ and there exists a 2-element $t \in C$ such that $\overline{t} \in \overline{C}\setminus F^*(\overline{C})$ and such that, up to permutation of $\{1, 2, 3\}$, the element $\overline{t}$ induces an inner automorphism on $\overline{E_3}$ and an outer automorphism of 2-power order on $\overline{E_1}$ and on $\overline{E_2}$. As $r(\overline{S}) \geq 4$, we may choose $\overline{t}$ to be an involution and such that $\overline{t}$ centralises an elementary abelian subgroup of $E(\overline{C})$ of order 8. Then $\overline{t}$ induces a field automorphism on $\overline{E_1}$ and on $\overline{E_2}$. Let $i \in \{1, 2\}$ and let $a_i, b_i \in E_i$ be elements of order 4 such that $a := a_1 a_2$ and $b := b_1 b_2$ are distinct commuting involutions that are centralised by t. Let $\widetilde{C} := \overline{C}/\langle \overline{z} \rangle$. Then $\widetilde{t}$ induces an inner automorphism on $\widetilde{E_3}$ and an involutory field automorphism on $\widetilde{E_1}$ and on $\widetilde{E_2}$. In particular we have that $C_{\widetilde{E_1}}(\widetilde{t})$ is a subfield subgroup and therefore $O(C_{\widetilde{E_1}}(\widetilde{t})) = 1$. By symmetry $O(C_{\widetilde{E_2}}(\widetilde{t})) = 1$. Set $A := \langle a, b, z, t \rangle$ and let $v \in A^\#$ be arbitrary. Then $O(C_{\widetilde{C}}(\widetilde{v}))$ induces inner automorphisms on $E(\widetilde{C})$ and thus $O(C_{\widetilde{C}}(\widetilde{v})) = O(C_{E(\widetilde{C})}(\widetilde{v}))$. Now let $V \leq A$ be of order 4 and recall Definition 5.1.

11. THE 2-RANK OF G

Our objective is to prove that $\Delta_V \leq C_C(A)O(C)$. If $z \in V$, then $\Delta_V \leq O(C)$, and if $t \in V$ or $tz \in V$, then $\widetilde{\Delta_V} \leq O(C_{E(\widetilde{C})}(\widetilde{t})) = O(C_{\widetilde{E_3}}(\widetilde{t})) = O(C_{\widetilde{E_3}}(\widetilde{A}))$ by the previous paragraph. If $a, b \in V$, then

$$\widetilde{\Delta_V} \leq O(C_{E(\widetilde{C})}(\widetilde{a})) \cap O(C_{E(\widetilde{C})}(\widetilde{b})) = O(\widetilde{E_3}) = 1$$

because, by the structure of the components of $\widetilde{C}$, the subgroups $C_{\widetilde{E_1}\widetilde{E_2}}(\widetilde{a})$ and $C_{\widetilde{E_1}\widetilde{E_2}}(\widetilde{b})$ intersect in a 2-group. This argument also yields that $\widetilde{\Delta_V} = 1$ if V contains a (or az) together with one of bz, ab, abz. By symmetry between a and b, the only case left to consider is the case where without loss $V = \langle a, bt \rangle$. Then we see that $\widetilde{\Delta_V} \leq O(C_{E(\widetilde{C})}(\widetilde{a}))$ which is a direct product of two cyclic groups of odd order that are inverted by $\widetilde{b}$ with a cyclic group of odd order that is centralised by $\widetilde{b}$. Let $i \in \{1, 2\}$. As $\widetilde{t}$ centralises a subfield subgroup of $\widetilde{E_i}$ that contains $\langle \widetilde{a_i}, \widetilde{b_i} \rangle$, it centralises a non-trivial subgroup of $O(C_{\widetilde{E_i}}(\widetilde{a}))$. Then it follows that $\widetilde{bt}$ inverts $O(C_{\widetilde{E_1}\widetilde{E_2}}(\widetilde{a}))$ and we deduce that

$$\widetilde{\Delta_V} \cap \widetilde{E_1}\widetilde{E_2} \leq O(C_{\widetilde{E_1}\widetilde{E_2}}(\widetilde{a})) \cap O(C_{\widetilde{E_1}\widetilde{E_2}}(\widetilde{bt})) = 1.$$

Therefore $\widetilde{\Delta_V} \leq O(C_{\widetilde{E_3}}(\widetilde{t})) = O(C_{\widetilde{E_3}}(\widetilde{A}))$ and hence $\widetilde{\Delta_V} \leq O(C_{\widetilde{C}}(\widetilde{A}))$. We conclude that $\overline{\Delta_V} \leq O(C_{\overline{C}}(\overline{A}))$ whence $\Delta_V \leq O(C)O(C_C(A))$ for all subgroups V of A of order 4.

As in Case 2 it follows that Θ (from Definition 5.1) defines a soluble A-signalizer functor, and this contradicts Lemma 5.6.

By Theorem **C** there are no more cases to consider, therefore it is impossible that $r_2(G) \geq 4$. $\square$

CHAPTER 12

The F*-Structure Theorem

As in the previous chapter, we begin with a series of lemmas, each of them handling a particular possibility for $F^*(\overline{C})$ (or at least for a component in $\overline{C}$). The bound on $r_2(G)$ from Theorem **D** is, of course, crucial. In many situations we can then still argue with signalizer functors or with the fact that there are no (weakly) balanced subgroups in C. All this information collected together yields Theorem 12.4, a list of possibilities regarding the number of components in $\overline{C}$ and their shape. What is left to be done for the F*-Structure Theorem is to go through these cases and to exclude a few more configurations where weakly balanced subgroups appear. The corresponding technical details are dealt with in Lemma 12.5.

The information from the F*-Structure Theorem is our starting point for some more local analysis in the next sections.

LEMMA 12.1. *Suppose that Hypothesis* 10.1 *holds and let* $E \in \mathcal{L}_2(C)$. *Then* $\overline{E} \not\simeq PSU_3(4)$.

PROOF. Assume otherwise and let $E \in \mathcal{L}_2(C)$ be such that $\overline{E} \simeq PSU_3(4)$. Then $\overline{E}$ is simple and of 2-rank 2, hence $\langle \overline{z} \rangle \overline{E}$ has 2-rank 3. Theorem **D** implies that $\overline{E}$ is the unique simple component of $\overline{C}$. Let $B \leq E$ be elementary abelian of order 4 and set $A := \langle z \rangle B$. Then B is the centre of a Sylow 2-subgroup of E (see for example page 30 in [**CCN103**]) and for all $b \in B^{\#}$ we have that $O(C_{\overline{E}}(\overline{b})) = 1$. With Lemma 5.15 we know that $O(C_{\overline{C}}(\overline{b}))$ centralises $O_2(\overline{C})$ and $E(C_{\overline{C}}(\overline{b}))$. The type of $\overline{E}$ (more precisely the fact that its outer automorphism group has order 4, again by [**CCN103**]) implies that $O(C_{\overline{C}}(\overline{b}))$ induces an inner automorphism on $\overline{E}$. It follows, for all $b \in B^{\#}$, that

$$O(C_{\overline{C}}(\overline{b})) = O(C_{\overline{E}}(\overline{b})) = 1$$

and then, for all $a \in A^{\#}$, that $O(C_C(a)) \leq O(C)$.

Thus A is balanced, which is a contradiction to Lemma 5.8. □

LEMMA 12.2. *Suppose that Hypothesis* 10.1 *holds, that* $E \in \mathcal{L}_2(C)$ *and that* q *is a power of an odd prime. Then* $\overline{E}$ *is not isomorphic to* $PSL_3(q)$ *or to* $PSU_3(q)$.

PROOF. Assume that $\overline{E} \simeq PSL_3(q)$ or $PSU_3(q)$. Then $\overline{E}$ is simple of 2-rank 2, so $\langle \overline{z} \rangle \overline{E}$ has 2-rank 3 whence by Theorem **D** it follows that $r_2(G) = 3$. This implies that $\overline{E}$ is the unique simple component.

Let $B \leq E$ be elementary abelian of order 4 and let $A := \langle z \rangle B$. For all $b \in B^{\#}$, we see that $C_{\overline{E}}(\overline{b})$ is a component of $C_{\overline{C}}(\overline{b})$ isomorphic to $SL_2(q)$ and therefore

$$[\overline{\alpha(b)}, C_{\overline{E}}(\overline{b})] \leq O(C_{\overline{C}}(\overline{b})) \cap E(C_{\overline{C}}(\overline{b})) = 1.$$

87

As $\overline{B} \leq C_{\overline{E}}(\overline{b})$, we conclude that $[\overline{\alpha(b)}, \overline{B}] = 1$. Then Lemma 5.5 (1) yields, for all $a \in A^\#$, that $\alpha(a) \leq O(C)O(C_C(A))$ and consequently A is a weakly balanced subgroup. This contradicts Lemma 5.8. $\square$

LEMMA 12.3. *Suppose that Hypothesis* 10.1 *holds and that* $E \in \mathcal{L}_2(C)$. *Then* $\overline{E}$ *is not isomorphic to* A_5, A_6, A_7, $PSL_2(7)$ *or* $PSL_2(9)$.

PROOF. Assume that $\overline{E}$ is isomorphic to one of the groups mentioned. We argue as for the previous lemmas – since $\overline{E}$ is simple of 2-rank 2, we have that $r_2(\langle \overline{z} \rangle \overline{E}) = 3$ and hence $r_2(G) = 3$ by Theorem **D**. Again this implies that $\overline{E}$ is the unique simple component.

We choose $B \leq E$ to be elementary abelian of order 4 such that $O(C_{\overline{E}}(\overline{B}))$ is cyclic of order 3 in the A_7-case and that $O(C_{\overline{E}}(\overline{B})) = 1$ in the other four cases. Let $A := \langle z \rangle B$. Then for all $b \in B^\#$, we know that

$$O(C_{\overline{E}}(\overline{b})) = O(C_{\overline{E}}(\overline{B})) = O(C_{\overline{E}}(\overline{A})).$$

Lemma 5.15 together with the fact that the outer automorphism group of $\overline{E}$ is a 2-group implies, for all $b \in B^\#$, that

$$O(C_{\overline{C}}(\overline{b})) = O(C_{\overline{E}}(\overline{b})) = O(C_{\overline{E}}(\overline{A}))$$

and therefore $\alpha(b) \leq O(C)O(C_C(A))$. With Lemma 5.5 (1) we conclude, for all $a \in A^\#$, that $\alpha(a) \leq O(C)O(C_C(A))$. Thus A is weakly balanced and this contradicts Lemma 5.8. $\square$

THEOREM 12.4. *Suppose that Hypothesis* 10.1 *holds. Then there exists a subgroup* $\overline{T}$ *of* $\overline{C}$ *such that* $F^*(\overline{C}) = E(\overline{C})\overline{T}$ *and* $\overline{T} = 1$ *or* $\overline{T} \simeq Q_8$. *Moreover either* $O^{2'}(\overline{C}) = F^*(\overline{C})$ *or* $|\mathcal{L}_2(C)| = 3$, *each member of* $\mathcal{L}_2(C)$ *is normal in* $O^{2'}(C)$ *and* C *is transitive on* $\mathcal{L}_2(C)$. *Finally, one of the following holds:*

(1) $\overline{T} \simeq Q_8$ *and there exists an odd number* $q \geq 11$ *such that* $E(\overline{C})$ *is isomorphic to* $PSL_2(q)$.

(2) $\overline{T} \simeq Q_8$ *and* $E(\overline{C})$ *is isomorphic to* $2A_7$ *or there exists an odd number* $q \geq 5$ *such that* $E(\overline{C}) \simeq SL_2(q)$.

(3) $\overline{T} = 1$ *and there exists an odd number* q *such that* $E(\overline{C})$ *is isomorphic to* $SL_4(q)$, $Sp_4(q)$ *or* $SU_4(q)$.

(4) $\overline{C}$ *has two components* $\overline{E_1}$ *and* $\overline{E_2}$ *and for all* $i \in \{1, 2\}$ *there exists an odd number* $q_i \geq 5$ *such that* $\overline{E_i} \simeq SL_2(q_i)$ *or* $\overline{E_i} \simeq 2A_7$.

(5) $\overline{C}$ *has two components* $\overline{E_1}$ *and* $\overline{E_2}$ *and there exist odd numbers* $q_1 \geq 11$ *and* $q_2 \geq 5$ *such that* $\overline{E_1} \simeq PSL_2(q_1)$ *and* $\overline{E_2} \simeq SL_2(q_2)$ *or* $\overline{E_2} \simeq 2A_7$. *Moreover* $\overline{T} = 1$.

(6) $\overline{T} = 1$ *and* $\overline{C}$ *has three components* $\overline{E_1}$, $\overline{E_2}$ *and* $\overline{E_3}$, *and for all* $i \in \{1, 2, 3\}$ *there exists an odd number* $q_i \geq 5$ *such that* $\overline{E_i} \simeq SL_2(q_i)$ *or* $\overline{E_i} \simeq 2A_7$.

PROOF. From Theorem **C** we know that $\overline{C}$ has one, two or three components. Hence we need to show that these three cases lead to (1)-(6) in the theorem. Let $S_0 \leq C$ be a 2-subgroup of C such that $\overline{S_0} = O_2(\overline{C})$. Then S_0 is of rank 1 by Theorem **B** and $r_2(C) \leq 3$ by Theorem **D**. Thus the hypothesis of Lemma 11.6 is satisfied for every member of $\mathcal{L}_2(C)$ and it follows that every component of $\overline{C}$ is isomorphic to a group from List (a) or List (b) in Definition 11.5.

Let $\widetilde{C} := \overline{C}/\langle \overline{z} \rangle$ and $L \in \mathcal{L}_2(C)$.

(i) If $\overline{L}$ is simple, then there exists an odd number $q \geq 11$ such that $\overline{L} \simeq PSL_2(q)$.

PROOF. Lemma 11.6 yields that $\overline{L}$ is isomorphic to one of the groups from List (b). Hence Lemmas 12.1, 12.2 and 12.3 imply the statement. □

(ii) If $\overline{L}$ is not simple, then either $r_2(L) = 1$ or there exists an odd number q such that $\overline{L} \simeq SL_4(q), Sp_4(q)$ or $SU_4(q)$ and $E(\overline{C}) = O^{2'}(\overline{C})$.

PROOF. It follows from Lemma 11.6 that $\overline{L}$ is isomorphic to a group from List (a). We suppose that $r_2(L) > 1$. Then there exists an odd number q such that $\overline{L} \simeq SL_4(q), Sp_4(q)$ or $SU_4(q)$. As $r_2(C) \leq 3$, this implies that $\overline{L}$ is the unique component of $\overline{C}$ and that $\overline{S_0}$ is not quaternion. The outer automorphism group of $\widetilde{L}$ is 2-nilpotent and $O^2(\widetilde{C}) = \widetilde{C}$ by Lemma 4.6, so $O^{2'}(\overline{C}) = \overline{L} = E(\overline{C})$. □

(iii) $\overline{L} \trianglelefteq O^{2'}(\overline{C})$.

PROOF. This follows because $|\mathcal{L}_2(C)| \leq 3$ by Theorem **C** and $O^2(\widetilde{C}) = \widetilde{C}$ by Lemma 4.6. □

(iv) Suppose that $|\mathcal{L}_2(C)| \leq 2$. Then $O^{2'}(\overline{C}) = E(\overline{C})\overline{S_0}$ and either $\overline{S_0} \simeq Q_8$ or $\overline{S_0} \leq E(\overline{C})$.

PROOF. The first statement follows from (iii) because by (i) and (ii), the outer automorphism group of every component of $\overline{C}$ is 2-nilpotent and $O^2(\widetilde{C}) = \widetilde{C}$ by Lemma 4.6. For the second statement we recall that $\overline{S_0}$ is cyclic or quaternion by Theorem **B** and hence the automorphism group of $\overline{S_0}$ is a 2-group unless $\overline{S_0} \simeq Q_8$. Thus $\overline{S_0} \simeq Q_8$ or $\overline{S_0} = \langle \overline{z} \rangle$ in which case Theorem **B**, part (i) and Lemma 11.1 yield that $\overline{S_0} \leq E(\overline{C})$. □

(v) If $\overline{L}$ is simple, then the theorem holds.

PROOF. Suppose that $\overline{L}$ is simple. Then by (i) there exists an odd number $q \geq 11$ such that $\overline{L} \simeq PSL_2(q)$ and (iv) yields that $\overline{S_0} \simeq Q_8$ or $\overline{S_0} \leq E(\overline{C})$. The first case leads to (1) and in the second case, the fact that $\overline{L}$ is simple implies that $\overline{C}$ has a component $\overline{E}$ distinct from $\overline{L}$. By Lemma 9.8 this means that $\overline{C}$ has exactly two components. Moreover $\overline{S_0} \leq \overline{E}$ whence $\overline{E}$ is not simple. It follows from (ii) and Theorem **D** that $r_2(\overline{E}) = 1$ and hence we have (5) by Lemma 9.5. □

(vi) If $E(\overline{C}) = \overline{L}$, then the theorem holds.

PROOF. By (v) we may suppose that $\overline{L}$ is not simple. Suppose that part (3) of the theorem does not hold. Then $r_2(L) = 1$ by (ii) whence there exists an odd number $q \geq 5$ such that $\overline{L}$ is isomorphic to $SL_2(q)$ or $2A_7$. Moreover (iv) yields that $\overline{S_0} \simeq Q_8$ or $\overline{S_0} \leq E(\overline{C})$. But if $\overline{S_0} \leq E(\overline{C})$, then C (and hence G) has quaternion Sylow 2-subgroups, contrary to Lemma 4.3 (1). Therefore, if (3) does not hold, then (2) holds. □

(vii) If $|\mathcal{L}_2(C)| = 3$, then the theorem holds.

PROOF. Lemma 9.8 implies (6). □

By (vi) and (vii) it suffices to consider the case where $\overline{C}$ has exactly two components. Thus let $E \in \mathcal{L}_2(C)$ be such that $E(\overline{C}) = \overline{LE}$. First suppose that one of the components is simple. Then (i) and Lemma 9.4 yield that, without loss, there exists an odd prime $q \geq 11$ such that $\overline{L} \simeq PSL_2(q)$ and $r_2(\overline{ES_0}) = 1$. Then (iv) and Lemma 9.5 imply that (5) holds. Next suppose that $\overline{L}$ and $\overline{E}$ are both not simple. Then Lemma 9.4 and (ii) give that both components have 2-rank 1. Thus (4) holds by Lemma 9.5.

This completes the proof of the theorem. □

LEMMA 12.5. *Suppose that Hypothesis 10.1 holds. Then $\overline{C}$ possesses at most one component of type A_7. Moreover if $\overline{C}$ has a component isomorphic to $2A_7$, then $O_2(\overline{C}) \not\simeq Q_8$.*

PROOF. For the first statement assume otherwise and let $E_1, E_2 \in \mathcal{L}_2(C)$ be such that $\overline{E_1}$ and $\overline{E_2}$ are of type A_7. By Lemma 12.3, both these components are not simple, hence they share the central involution $\overline{z}$ and we are in case (4) or (6) of Theorem 12.4. By Lemma 2.13 we may choose an elementary abelian subgroup $B \leq E_1 E_2$ such that $z \notin B$. Moreover we choose B such that $\overline{B}$ centralises a subgroup of order 3 of $\overline{E_1}$ and of $\overline{E_2}$. Set $A := B\langle z \rangle$. We show that A is weakly balanced:

By choice of B, we have for all $b \in B^{\#}$ and $i = 1, 2$ that the groups $O(C_{\overline{E_i}}(\overline{b}))$ are cyclic of order 3 and centralise all of $\overline{A}$. Hence, for all $a \in A^{\#}$, it follows that

$$O(C_{E_1}(a))O(C_{E_2}(a)) \leq O(C)O(C_C(A)).$$

Let $b \in B^{\#}$. Then $\overline{\alpha(b)}$ centralises $O_2(\overline{C})$ and, if it exists, the third component of $\overline{C}$, by Lemma 5.15. Also, since it has odd order, the subgroup $\overline{\alpha(b)}$ induces inner automorphisms on $\overline{E_1}$ and $\overline{E_2}$. Thus $\overline{\alpha(b)} \leq F^*(\overline{C})$ and we deduce that

$$\overline{\alpha(b)} \leq O(C_{F^*(\overline{C})}(\overline{b})) = O(C_{F^*(\overline{C})}(\overline{A})).$$

This implies that $\alpha(b) \leq O(C)O(C_C(A))$. With Lemma 5.5 (1) it follows, for all $a \in A^{\#}$, that $\alpha(a) \leq O(C)O(C_C(A))$. But then A is weakly balanced, which contradicts Lemma 5.8.

For the second statement suppose that $\overline{E}$ is a component of $\overline{C}$ isomorphic to $2A_7$ and assume that $O_2(\overline{C}) \simeq Q_8$. Let $T \in \mathrm{Syl}_2(O_{2',2}(C))$. Then by Lemma 2.13 we may choose an elementary abelian subgroup $B \leq ET$ such that $z \notin B$, and again we choose B such that $\overline{B}$ centralises a subgroup of order 3 of $\overline{E}$. Then $A := B\langle z \rangle$ is weakly balanced just as in the previous paragraph. □

THEOREM 12.6 (The F*-Structure Theorem). *Suppose that Hypothesis* 10.1 *holds. Then there exists an odd number q (or odd numbers q_1, q_2, q_3) such that $F^*(\overline{C})$ is isomorphic to one of the following groups:*

List I
- $Q_8 * SL_2(q)$;
- $SL_2(q_1) * SL_2(q_2)$;
- $Q_8 * SL_2(q_1) * SL_2(q_2)$;
- $SL_2(q) * 2A_7$.

All these products have a common central involution.

List II
- $Sp_4(q)$;
- $SL_4(q)$;
- $SU_4(q)$.

List III
- $SL_2(q_1) * SL_2(q_2) * 2A_7$;
- $SL_2(q_1) * SL_2(q_2) * SL_2(q_3)$.

These products have a common central involution.

List IV
- $Q_8 \times PSL_2(q)$ and $q \geq 11$;
- $2A_7 \times PSL_2(q)$ and $q \geq 11$;
- $SL_2(q_1) \times PSL_2(q_2)$ and $q_2 \geq 11$.

PROOF. We go through the cases in Theorem 12.4.

First (1) gives the first case on List IV and (2) gives the first case on List I. Then (3) gives List II. Looking at (4), we see that the only remaining cases are those on List I, by Lemma 12.5. We turn to (5) and obtain precisely the last two cases on List IV. Finally (6) and Lemma 12.5 give List III.

Although this is clear from Theorem **B**, we mention at this point that whenever the products describing the shape of $F^*(\overline{C})$ are not direct, then there is a unique central involution in the intersection of the factors and this involution is $\overline{z}$. □

CHAPTER 13

More Involutions

Our starting point is a series of statements following from the F*-Structure Theorem. Lemmas 13.4 and 13.5 are at the heart of the analysis – there we use our information about the 2-structure of C in order to deduce knowledge about the centralisers of specially chosen involutions. This leads to more maximal subgroups H of G containing the centraliser of an involution and with this involution being isolated in H. To make our life easier, we refer to particular cases in the F*-Structure Theorem by saying that "$F^*(\overline{C})$ **is as on List I, II, III or IV**", respectively.

13.1. Preliminary Results

LEMMA 13.1. *Suppose that Hypothesis* 10.1 *holds and let* $t \in C$ *be an involution distinct from* z. *Then the following hold:*

(1) G *is simple.*
(2) $O^2(C) = C$.
(3) *Either* $F^*(\overline{C}) = O^{2'}(\overline{C})$ *or every member of* $\mathcal{L}_2(C)$ *is normal in* $O^{2'}(C)$, *and* $|\mathcal{L}_2(C)| = 3$ *with* C *acting transitively on* $\mathcal{L}_2(C)$.
(4) $G = \langle C_G(t), C_G(tz) \rangle$.
(5) $C_G(t)$, $C_G(tz)$ *and* C *are pairwise distinct. If* $C \le M \max G$, $C_G(t) \le H_t \max G$ *and* $C_G(tz) \le H_{tz} \max G$, *then* M, H_t *and* H_{tz} *are pairwise distinct.*

PROOF. Let $S \in \mathrm{Syl}_2(C)$. The F*-Structure Theorem 12.6 implies that $z \in \Phi(S)$, so (1) and (2) follow from Lemmas 3.1 (10), 4.4 and 4.5. Theorem 12.4 yields (3), and Theorem 3.6, Lemma 4.7 and (1) imply that (4) holds. From there we deduce that $C_G(t)$ and $C_G(tz)$ are distinct and that H_t and H_{tz} are distinct, because these subgroups contain $C_K(t)$ and $C_K(tz)$, respectively. The involution centralisers $C_G(t)$ and $C_G(tz)$ are not equal to C by Lemma 4.13. For the same reason, the subgroups H_t and H_{tz} are distinct from M. Thus (5) is proved as well. □

LEMMA 13.2. *Suppose that Hypothesis* 10.1 *holds, that* $O^{2'}(\overline{C}) \neq F^*(\overline{C})$ *and that* $\mathcal{L}_2(C) = \{E_1, E_2, E_3\}$. *Let* $q \ge 5$ *be an odd number such that* $\overline{E_i} \simeq SL_2(q)$ *for all* $i \in \{1, 2, 3\}$ *and suppose that* $t \in C \backslash L(C)$ *is an involution. Then* $\overline{t}$ *induces, up to permutation of* $\{1, 2, 3\}$, *an outer automorphism in* $PGL_2(q)$ *on* $\overline{E_1}$ *and* $\overline{E_2}$ *and centralises* $\overline{E_3}$.

PROOF. Let $E := L(C)$ and $\widetilde{C} := \overline{C}/\langle \overline{z} \rangle$. Then $\widetilde{E}$ is a direct product of three groups isomorphic to $PSL_2(q)$ and therefore $\mathrm{Aut}(\widetilde{E})$ is the wreath product of $\mathrm{Aut}(\widetilde{E_1})$ with $\mathcal{S}_3$. For all $i \in \{1,2,3\}$ we know from Lemma 13.1 (3) that $E_i \trianglelefteq O^{2'}(C)$ and hence $\widetilde{C}/C_{\widetilde{C}}(\widetilde{E})$ is the semi-direct product of $\widetilde{E}$ with a subgroup isomorphic to A_4. (Recall that $O^2(\widetilde{C}) = \widetilde{C}$ by Lemma 4.6.) Moreover, up to a permutation of $\{1,2,3\}$, the involution $\widetilde{t}$ acts as an involutory outer automorphism on $\widetilde{E_1}$ and $\widetilde{E_2}$ and induces an inner automorphism on $\widetilde{E_3}$. Suppose that $\widetilde{t}$ induces a field automorphism on $\widetilde{E_1}$. Then it induces a field automorphism on $\widetilde{E_2}$ as well and so there exists a prime power q_0 dividing q such that $C_{\overline{E_1 E_2}}(\overline{t})$ contains a subgroup isomorphic to $SL_2(q_0) * SL_2(q_0)$. This subgroup has 2-rank 3 by Lemma 2.13 and $\overline{t}$ centralises it, but is not contained in it. Then $r_2(C) \geq 4$ contrary to Theorem **D**. By symmetry it follows that $\widetilde{t}$ induces an outer automorphism in $PGL_2(q)$ on $\overline{E_1}$ and $\overline{E_2}$.

Next let $t \in P \in \mathrm{Syl}_2(C)$ be such that $T := C_P(t)$ is a Sylow 2-subgroup of $C_C(t)$. For all $i \in \{1,2,3\}$ set $P_i := P \cap E_i$. Then the previous paragraph implies that there are elements $u_1 \in P_1$ and $u_2 \in P_2$ of order 4 that are inverted by t and such that $C_{P_1 P_2}(t) = \langle u_1 u_2, z \rangle$. Assume that t does not centralise P_3. We know from the previous paragraph that $\overline{t}$ induces an inner automorphism on $\overline{E_3}$ and hence there exists an element $u_3 \in P_3$ of order 4 such that u_3 is inverted by t. As $u_1 u_2$ and $u_1 u_3$ are involutions, the subgroup $\langle t, u_1 u_2, u_1 u_3, z \rangle$ is elementary abelian of order 16, contrary to Theorem **D**. Thus $\overline{t}$ induces an inner automorphism on $\overline{E_3}$ that centralises $\overline{P_3}$ and it follows that $\overline{t}$ centralises $\overline{E_3}$. $\square$

LEMMA 13.3. *Suppose that Hypothesis 10.1 holds. Let $t \in C$ be an involution such that $t \neq z$ and suppose that B is an elementary abelian subgroup of order 4 of C that contains t, but not z. Let $H < G$.*

(1) *There exists an involution $b \in B$ such that $C_G(b) \not\leq H$. In particular, if $C_G(t) \leq H$, then $N_H(B)/C_H(B)$ is not transitive on $B^\#$.*

(2) *Suppose that $F^*(\overline{C})$ is as on List IV and that $E \in \mathcal{L}_2(C)$ is such that $\overline{E}$ is simple and $B \leq E$. Suppose that $C_G(t) \leq H$. Then there exists an element $b \in B^\#$ such that $C_C(b) \not\leq H$.*

PROOF. Let t, b and tb be the involutions in B. For (1) we assume that $C_G(t)$, $C_G(b)$ and $C_G(tb)$ are all contained in H. Let H_{tz} denote a maximal subgroup of G containing $C_G(tz)$. Then $B \leq C_G(tz)$ and therefore B acts coprimely on $O(H_{tz})$. With Lemma 2.1 (4) this yields that

$$O(H_{tz}) = \langle C_{O(H_{tz})}(v) \mid v \in B^\# \rangle \leq \langle C_G(t), C_G(b), C_G(tb) \rangle \leq H.$$

As $z \in H < G$, Lemma 4.2 (2) implies that $C_K(t) \subseteq O(H)$. Similarly $C_K(tz) \subseteq O(H_{tz})$ and thus with Lemma 13.1 (4) it follows that

$$G = \langle C_K(t), C_K(tz) \rangle \leq \langle O(H), O(H_{tz}) \rangle \leq H.$$

This is impossible. In particular $N_H(B)/C_H(B)$ is not transitive on $B^\#$.

We move to (2) and suppose that one of the cases from List IV holds and that $E \in \mathcal{L}_2(C)$ is such that $\overline{E}$ is simple and $B \leq E$.

Assume that $C_C(b)$ and $C_C(bt)$ are subgroups of H. Then H contains $C_G(t)$, $C_C(b)$ and $C_C(bt)$ and the coprime action of B on $O(C)$, together with Lemma 2.1 (4), implies that $O(C) \leq H$. By hypothesis there exists an odd number $q \geq 11$

such that $\overline{E} \simeq PSL_2(q)$. As $q \geq 11$, the subgroup structure of $\overline{E}$ (see for example Dickson's Theorem 6.5.1 in [**GLS98**]) gives that $\overline{E} = \langle C_{\overline{E}}(\overline{t}), C_{\overline{E}}(\overline{b}) \rangle$. Therefore

$$EO(C) = \langle C_E(t), C_E(b) \rangle O(C) \leq H.$$

It follows that all involutions in E are not only contained in H, but also conjugate in H. We deduce that, for all involutions $s \in C$ distinct from z, the centraliser $C_C(s)$ is contained in H (because s itself or sz is conjugate to t in $C_H(z)$). If $\overline{E}$ is the only component in $\overline{C}$, then it follows that $O_{2',2}(C) \leq C_C(t)O(C) \leq H$. If a second component $\overline{E_2}$ exists, then $E_2 \leq C_C(t)O(C) \leq H$. In both cases we have that $O_{2',F^*}(C) \leq H$. We conclude that $C \leq H$:

For an arbitrary element $c \in C \backslash O_{2',F^*}(C)$, we know that $\overline{c}$ centralises $\overline{E}$ or induces a non-trivial outer automorphism on it, and in the former case $c \in C_C(t)O(C) \leq H$ because $\overline{c} \in C_{\overline{C}}(\overline{t})$. Our hypothesis that $F^*(\overline{C})$ is as on List IV yields that $|\mathcal{L}_2(C)| \leq 2$ and thus $\overline{C}/F^*(\overline{C})$ has odd order by Lemma 13.1 (3). Therefore we may suppose that $\overline{c}$ has odd order. An outer automorphism of odd order of $\overline{E}$ must be a field automorphism and hence centralises an involution in $\overline{E}$. Consequently there exists an involution $s \in C$ such that $\overline{c} \in C_{\overline{C}}(\overline{s})$ which means that $c \in C_C(s)O(C) \leq H$. Thus we showed that $C \leq H$. As $H < G$, this means that G has a maximal subgroup that contains C and $C_G(t)$, contrary to Lemma 4.13. $\square$

LEMMA 13.4. *Suppose that Hypothesis* 10.1 *holds and that* $a \in C$ *is an involution distinct from* z *and chosen as follows:* $\overline{a} \in F^*(\overline{C})$; *if* $r_2(E(\overline{C})) \geq 2$, *then* $\overline{a} \in E(\overline{C})$; *if* $\overline{C}$ *has a simple component* $\overline{L}$, *then* $\overline{a} \in \overline{L}$.

Let $v \in \{a, az\}$ and let H be a maximal subgroup of G containing $C_G(v)$. Then
(1) v is isolated in H and
(2) either $vz \in v^C$ or $F^*(\overline{U})$ is as on List IV.

PROOF. Let $S_0 \in \text{Syl}_2(C_C(v))$ and $S_0 \leq S \in \text{Syl}_2(C)$.

(i) Either $|S : S_0| = 2$ and $vz \in v^S$ or $F^*(\overline{C})$ is as on List IV and $S_0 = S$. In particular (2) holds.

PROOF. Is is immediate from the groups on List IV that, if $F^*(\overline{C})$ is as on List IV, then $S_0 = S$.

Now suppose that $F^*(\overline{C})$ is as on List II and let $E \in \mathcal{L}_2(C)$. Let q be a power of an odd prime and let U be the 4-dimensional module over a field of order q defining $\overline{E}$. Let U_1, U_2 be the eigenspaces of $\overline{a}$. As we saw in the proof of Theorem **D**, there are subgroups $\overline{L_1}$ and $\overline{L_2}$ of $C_{\overline{E_2}}(\overline{a})$ such that $\overline{L_1} \times \overline{L_2} \trianglelefteq C_{\overline{E_2}}(\overline{a})$ and such that, for all $i \in \{1, 2\}$, the subgroup $\overline{L_i}$ acts faithfully as $SL_2(q')$ on U_i and centralises U_{3-i}. All involutions in $\overline{E}$ distinct from $\overline{z}$ have 2-dimensional eigenspaces on U and they are all conjugate in $\overline{E}$. In particular $az \in a^S$ in this case and therefore $vz \in v^S$. Since there exists an element in S interchanging a and az, we see that $|S : S_0| = 2$. Hence (i) holds in this case.

Finally we suppose that $F^*(\overline{C})$ is as on List I or III. Then all components of $\overline{C}$ have quaternion Sylow 2-subgroups and there exist $E_1, E_2 \in \mathcal{L}_2(C)$ such that $\overline{v}$ is diagonal in $\overline{E_1} * \overline{E_2}$. So (i) follows because $\overline{E_1}$ and $\overline{E_2}$ are normal in $O^{2'}(\overline{C})$. $\square$

(ii) $vz \notin v^H$.

PROOF. This follows from Lemma 13.1 (4) because $C_K(v) \subseteq H$. □

(iii) $S_0 \in \mathrm{Syl}_2(H)$.

PROOF. We know that $S \in \mathrm{Syl}_2(G)$ and therefore the statement follows from (i) and (ii). □

From now on we assume that v is not isolated in H. Then there exists an element $h \in H$ such that $v^h \in S_0$, but $v^h \neq v$. It follows from Alperin's Fusion Theorem (see for example (38.1) in [**Asc00**], applied to H, S_0, v and v^h) that S_0 has a subgroup R such that $v \in R$ and $N_{S_0}(R) \in \mathrm{Syl}_2(N_H(R))$, but $N_H(R) \not\leq C_G(v)$.

We set $B := \langle v^{N_H(R)} \rangle$ and we denote by A the group of automorphisms of B induced by $N_H(R)$.

(iv) B is elementary abelian of order 8. In particular $z \in B$.

PROOF. As $v \in Z(S_0)$ and $v \in R \leq S_0$, we have that $v \in Z(R)$. Thus $B \leq Z(R)$ and B is elementary abelian. The rank of B is 2 or 3 by Theorem **D** and because $N_H(R)$ does not centralise v. The property that $N_{S_0}(R) \in \mathrm{Syl}_2(N_H(R))$ implies that A is not a 2-group, hence if B is a fours group, then A acts transitively on $B^\#$. This contradicts Lemma 13.3 (1) and consequently B has order 8. As z centralises B, Theorem **D** implies that $z \in B$. □

(v) A is cyclic of order 3 or isomorphic to $\mathcal{S}_3$. Moreover $B = \langle z \rangle \times [B, A]$.

PROOF. From (iv) we know that $\mathrm{Aut}(B) \simeq GL_3(2)$. As z is isolated in H, we see that A is isomorphic to a subgroup of an involution centraliser in $GL_3(2)$. In particular A is a $\{2,3\}$-group. Moreover the group of automorphisms induced by $N_{S_0}(R)$ on B is a Sylow 2-subgroup of A (and centralises a), so it follows that $|v^A|$ is odd. This yields the first result and Lemma 2.1 (2), applied to an element of order 3 in A, yields the second statement. □

(vi) $v \notin [B, A]$. In particular $B = \langle v \rangle [B, A]$.

PROOF. Otherwise, as A acts transitively on $[B, A]$, we have a contradiction to Lemma 13.3 (1). □

(vii) $\overline{B} \leq F^*(\overline{C})$.

PROOF. We have that $z \in B \leq R$ and therefore $N_H(R) \leq C$. Moreover $\overline{a} \in F^*(\overline{C})$ by hypothesis and $\overline{z} \in O_2(\overline{C})$, so $\overline{v} \in F^*(\overline{C})$. As $\overline{C}$ controls fusion in $\overline{C}$ by Lemma 3.1 (9), it follows that $\overline{B} = \langle \overline{v}^{N_H(R)} \rangle \leq F^*(\overline{C})$. □

(viii) $F^*(\overline{C})$ is as on List IV and $\overline{a}$ is contained in the simple component of $\overline{C}$.

PROOF. If $F^*(\overline{C})$ is as on List IV, then this follows from the choice of a in the hypothesis of the lemma. So we assume otherwise and let $d \in [B, A]$ be an involution. Then $d \in B \leq S_0$, but $d \neq z$ and $d \neq v$ by (v) and (vi). Then the structure of S in the cases from Lists I-III implies that d is conjugate to dz in S_0. It follows with (v) that all involutions in $B \backslash \langle z \rangle$ are conjugate in H, contrary to (ii). □

In light of (viii) let $L \in \mathcal{L}_2(C)$ be such that $\overline{L}$ is simple and $a \in L$. By (v) there exists an element $x \in A$ of order 3.

First suppose that $v = a$. Then $[a, x] = [B, A]$ and, as $x \in C$ and $a \in L \trianglelefteq C$, we deduce that $[a, x] \in L$. Therefore $[B, A] \leq L$. Then (vi) yields that $B = \langle a \rangle [B, A] \leq L$, which is impossible because $z \notin L$.

Next suppose that $v = az$. Then $[az, x] = [B, A]$ and hence $[B, A]$ contains an involution d that is conjugate to a in L. But $N_H(R) \leq C$ and $L \trianglelefteq C$, so it follows that $[B, A] \leq \langle d^{N_H(R)} \rangle \leq L$. This is a contradiction because $az \notin L$.

This completes the proof. $\square$

LEMMA 13.5. *Suppose that Hypothesis 10.1 holds and that a is chosen as in Lemma 13.4. Let $v \in \{a, az\}$, let H be a maximal subgroup of G containing $C_G(v)$ and set $V := \langle a, z \rangle$ and $\widehat{H} := H/O(H)$. Then the following hold:*

(1) *All involutions in V are isolated in H.*

(2) *Either v and vz are conjugate or $F^*(\overline{C})$ is as on List IV.*

(3) *If $F^*(\overline{C})$ is as on List I, then H is soluble.*

(4) *If $F^*(\overline{C})$ is as on List II, then either $\widehat{H}$ is soluble (in which case $q = 3$) or $q \neq 3$ and $\widehat{H}$ has two components isomorphic to $SL_2(q)$ with $\widehat{a}$ and $\widehat{az}$ being their central involutions, respectively.*

(5) *If $F^*(\overline{C})$ is as on List III, then $\widehat{H}$ has precisely one component, its 2-rank is 1 and $\widehat{z}$ is its central involution.*

(6) *If $F^*(\overline{C})$ is as on List IV, then either $\mathcal{L}_2(H) = \varnothing$ or $\widehat{H}$ has precisely one component, its 2-rank is 1 and $\widehat{z}$ is its central involution.*

(7) *V centralises $O_2(H)$ and $E(H)$.*

(8) *For all primes r we have that $\mathsf{M}^*_H(V, r) \subseteq Syl_r(H)$.*

PROOF. As z is isolated in H, Lemma 13.4 yields (1) and (2).

Suppose that $F^*(\overline{C})$ is as on List I. Then $C_{\overline{C}}(\overline{a})$ is soluble and therefore $C_C(v)$ is soluble. Lemma 4.2 (1) yields that $C_G(v) = C_C(v)O(C_G(v))$ is soluble and then the same result, applied to H and v, gives that H is soluble. This proves (3).

In (4)-(6), if $\mathcal{L}_2(H) \neq \varnothing$, then we always let $L \in \mathcal{L}_2(H)$ and $L_0 := O^\infty(C_L(V))$. Before we turn to (4), we also observe the following:

As $V \leq Z^*(H)$ by (1), we have that $\widehat{L_0} = \widehat{L}$. Moreover $L_0 O(C_G(v)) \in \mathcal{L}_2(C_G(v))$ whence Lemma 9.3 forces $\overline{L_0}$ to be contained in a component of $\overline{C}$. This implies that $\overline{L_0} \leq C_{E(\overline{C})}(\overline{v})$ and therefore $\overline{L_0}$ is one of the components of $C_{\overline{C}}(\overline{v})$.

In (4) we suppose that $F^*(\overline{C})$ is as on List II and we let q denote the order of the field that appears in the type of the unique component of $\overline{E}$ of $\overline{C}$. Then $C_{\overline{E}}(\overline{v})$ (and hence $C_{\overline{C}}(\overline{v})$) is soluble if an only if $q = 3$. If $q \geq 5$, then $C_{\overline{C}}(\overline{v})$ has two components isomorphic to $SL_2(q)$ that have central involutions $\overline{v}$ and $\overline{vz}$, respectively. In this case $\overline{L_0} \leq C_{\overline{E}}(\overline{v})$ by our observation and therefore $\overline{L_0}$ is one of the components of $C_{\overline{E}}(\overline{v})$. As $C_C(v) \leq H$, it follows that $\widehat{H}$ has exactly two components and that they are both isomorphic to $SL_2(q)$, with central involutions $\widehat{a}$ and $\widehat{az}$. Thus (4) is proved.

Suppose that $F^*(\overline{C})$ is as on List III and let $E_1, E_2, E_3 \in \mathcal{L}_2(C)$ be such that $\overline{a} \in \overline{E_1 E_2}$. Then $\overline{E_3}$ is the unique component of $C_{\overline{C}}(\overline{v})$ and, since $\overline{L_0}$ centralises $\overline{V}$, our observation implies that $\overline{L_0} \leq \overline{E_3}$. As $C_{E_3}(v) \leq H$, it follows that $\overline{L_0} = \overline{E_3}$ and therefore $\widehat{L}$ is the unique component of $\widehat{H}$, its 2-rank is 1 and $\widehat{z}$ is its central involution. This yields (5).

Suppose that $F^*(\overline{C})$ is as on List IV. If the first case from List IV holds, then $C_C(v)$ is soluble and hence (1) and Lemma 4.2 (1) imply that H is soluble. Thus $\mathcal{L}_2(H) = \varnothing$ in this case. Now suppose that one of the other two cases from List IV holds and let $E_1, E_2 \in \mathcal{L}_2(C)$ be such that $\overline{E_1}$ is simple. Then $\overline{L_0} \leq \overline{E_2}$ by our observation and because $\overline{L_0}$ centralises $\overline{C}$. Conversely $C_{E_2}(v) \leq H$ and therefore $\widehat{H}$ has a unique component, it is isomorphic to $\overline{E_2}$ and therefore of 2-rank 1, and its central involution is $\widehat{z}$. Hence (6) is proved.

Lemma 4.2 (5), applied to H and its isolated involutions z and v yields (7). Finally Lemma 3.11, applied to H and V, implies that there exist V-invariant Sylow r-subgroups of H for all primes r. □

HYPOTHESIS 13.6.
In addition to Hypothesis 10.1, *suppose the following:*
- *a is an involution in C distinct from z that is chosen as in Lemma* 13.4 *and $V := \langle z, a \rangle$.*
- *We let $\pi := \pi(F(M))$ and if possible, we choose M such that there is some $p \in \pi$ with $C_{O_p(M)}(z) = 1$.*
- *If $v \in \{a, az\}$, then H_v denotes a maximal subgroup of G such that $C_G(v) \leq H_v$. We set $\pi_v := \pi(F(H_v))$ and choose H_v such that, if possible, there exists a prime $p \in \pi(F(H_v))$ with $C_{O_p(H_v)}(v) = 1$. Moreover let T_v be a Sylow 2-subgroup of H_v with $V \leq Z(T_v)$ and $T_v \leq S \in Syl_2(C)$. We abbreviate $\widehat{H_v} := H_v / O(H_v)$.*
- *We choose H_a and H_{az} to be conjugate if a and az are conjugate in G.*
- *For all $v \in \{a, az\}$, if $C_G(v) \neq H_v$, then let $r_v \in \pi_v$ be such that $O_{r_v}(H_v)$ contains a v-minimal subgroup U_v. If $C < M$, then let $p \in \pi$ be such that $O_p(M)$ contains a z-minimal subgroup U.*

LEMMA 13.7. *Suppose that Hypothesis* 13.6 *holds, let $v \in \{a, az\}$ and suppose that $t \in v^C$. Then Hypothesis* 6.6 *is satisfied by t, H_t and U_t.*

PROOF. By hypothesis t is an involution in C. If $C_G(t) \leq H \max G$, then t is isolated in H by Lemma 13.5 (1), because t is conjugate to v. The remainder of Hypothesis 6.6 is notation and follows from Hypothesis 13.6. □

LEMMA 13.8. *Suppose that Hypothesis* 13.6 *holds, let $v \in \{a, az\}$ and suppose that $E(H_v) \neq 1$. Then $F^*(\overline{C})$ is as on List II, there is a unique component L in H_v, there exists an odd number $q \geq 5$ such that $L \simeq SL_2(q)$, and $v \in L$.*

PROOF. Lemma 13.1 (5) yields that $H_v \not\leq C$, therefore Lemma 4.2 (6) implies that $[F(H_v), z] \neq 1$. In particular $z \notin E(H_v)$. Let L be a component of H_v. Then it follows from Lemma 13.5 (3)-(6) that $F^*(\overline{C})$ is as on List II, that there exists an odd number $q \geq 5$ such that $L \simeq SL_2(q)$ and that L contains v or vz as central involution. If $Z(L) = \langle vz \rangle$, then, as v and vz are not conjugate in H_v, it follows that $vz \in Z(H_v)$ contrary to Lemma 13.1 (5). Thus $Z(L) = \langle v \rangle$ and in particular L is the unique component. □

LEMMA 13.9. *Suppose that Hypothesis* 13.6 *holds and that* $C \neq M$. *Then* M *has odd prime characteristic or* M *has a unique component* E. *In the latter case, the component* E *is simple and* $F^*(\overline{C})$ *is as on List IV. In particular we can choose the involution* a *to be contained* E.

PROOF. If M is not of odd prime characteristic, then Theorem **A** yields that $E(M) \neq 1$. Hence assume that E is a component of M. Then $E \leq C$ by Lemma 4.2 (5) and therefore E is a component of C. From the F*-Structure Theorem 12.6 we know the possibilities for $\overline{E}$ and therefore inspection of the Schur multipliers of the groups appearing in the lists yields that E is either simple or contains a central involution. (Note that $\overline{E} \not\simeq PSL_2(9)$ or A_7 by the F*-Structure Theorem, therefore $E \not\simeq 3PSL_2(9)$ or $3A_7$.) Theorem **B** implies that the only involution that can be central in a component of C is z. However, if $z \in E$, then we see a contradiction to Lemma 4.2 (6), applied to M. Thus we have that $z \notin E$ and in particular E is simple. Then one of the cases from List IV occurs, more precisely there exists an odd number $q \geq 11$ such that $E \simeq PSL_2(q)$. The last statement follows from the choice of a in Lemma 13.4. □

LEMMA 13.10. *Suppose that Hypothesis* 13.6 *holds. Let* $v \in \{a, az\}$ *and suppose that* $C_G(v)$ *is a maximal subgroup and that* a *and* az *are not conjugate. Suppose that* $d \in C$ *is distinct from* v, *but centralises* v *and is conjugate to* v *in* C. *Set* $H_d := C_G(d)$. *Then* $H_v \not\hookrightarrow H_d$ *and vice versa. In particular* $H_v \neq H_d$.

PROOF. By symmetry between d and v we may assume that $H_v \hookrightarrow H_d$. As a and az are not conjugate by hypothesis, Lemma 13.8 yields that $E(H_v) = 1 = E(H_d)$. Then the Infection Theorem (5) gives that $H_v = H_d$. In particular $\langle v, d \rangle \leq Z(H_v)$ and $H_v = C_G(vd)$ because H_v is primitive by Corollary 4.8. The fours group $\langle v, d \rangle$ centralises z and hence vz. If we set $w := vz$, then Lemma 2.1 (4) implies that
$$O(C_G(w)) = \langle C_{O(C_G(w))}(v), C_{O(C_G(w))}(d), C_{O(C_G(w))}(vd) \rangle \leq H_v.$$
In particular $C_K(w)$ is contained in H_v and this contradicts Lemma 13.1 (4). □

LEMMA 13.11. *Suppose that Hypothesis* 13.6 *holds, let* $v \in \{a, az\}$ *and suppose that* $t \in C$ *is an involution that is conjugate to* v. *Let* $\overline{W}$ *be a nilpotent* $C_{F^*(\overline{C})}(\bar{t})$-*invariant* $2'$-*subgroup of* $F^*(\overline{C})$. *Then* $[\overline{W}, \bar{t}] = 1$.

PROOF. First we note that $\overline{W} \leq E(\overline{C})$ because $\overline{W}$ has odd order, and for all components $\overline{E}$ of $\overline{C}$, the subgroup $\overline{W} \cap \overline{E}$ is $C_{\overline{E}}(\bar{t})$-invariant. Moreover $\bar{t} \in F^*(\overline{C})$ by hypothesis and Lemma 3.1 (9), because $F^*(\overline{C}) \trianglelefteq \overline{C}$. Let $\overline{E}$ be a component of $\overline{C}$ and let $\widetilde{C} := \overline{C}/\langle \bar{z} \rangle$.

First suppose that $F^*(\overline{C})$ is as on List II. Then $\bar{t} \in \overline{E}$ because $\bar{v} \in \overline{E}$ and v and t are C-conjugate. Moreover $C_{\overline{E}}(\bar{t})$ has index 2 in a maximal subgroup of $\overline{E}$. As $\overline{W}$ has odd order and lies in $\overline{E}$, it follows that $\overline{W} \leq C_{\overline{E}}(\bar{t})$ and hence $[\overline{W}, \bar{t}] = 1$.

Next suppose that $F^*(\overline{C})$ is as on List IV and recall that the simple component of $\overline{C}$ contains $\overline{v}$ or $\overline{vz}$ and hence it contains $\overline{t}$ or $\overline{tz}$. If $\overline{E}$ is not simple, then it therefore centralises $\overline{t}$, consequently $\overline{E}$ normalises $\overline{W}$ and hence it normalises the projection of $\overline{W}$ on $\overline{E}$. As $\overline{W}$ has odd order, its projection on $\overline{E}$ is now trivial and therefore $\overline{W}$ is contained in the simple component of $\overline{C}$. Hence suppose that $\overline{E}$ is simple. The type of $\overline{E}$ implies that $C_{\widetilde{E}}(\widetilde{t})$ is a dihedral group with central involution $\widetilde{t}$ or $\widetilde{t}z$, and this is a maximal subgroup of $\widetilde{E}$, therefore $\widetilde{W} \leq C_{\widetilde{E}}(\widetilde{t})$ and again $[\overline{W}, \overline{t}] = 1$.

We are left with the case that $F^*(\overline{C})$ is as on List I or III. If $\widetilde{t}$ centralises $\widetilde{E}$, then $\widetilde{W}$ centralises $\widetilde{E}$ because $\widetilde{W} = [\widetilde{W}, \widetilde{t}]$ and because $\widetilde{W}$ normalises $\widetilde{E}$. In particular the projection of $\widetilde{W}$ on every component that is centralised by $\widetilde{t}$ is trivial. If $\widetilde{t}$ does not centralise $\widetilde{E}$, then there exists an involution $\widetilde{u} \in \widetilde{E}$ such that $C_{\widetilde{E}}(\widetilde{t}) = C_{\widetilde{E}}(\widetilde{u})$. From the type of $\overline{E}$ it follows that this is a dihedral subgroup with central involution $\widetilde{u}$ and it is a maximal subgroup of $\widetilde{E}$. We conclude that the projection of $\widetilde{W}$ on $\widetilde{E}$ is contained in $C_{\widetilde{E}}(\widetilde{t})$ and hence $[\widetilde{W}, \widetilde{t}] = 1$. Thus $[\overline{W}, \overline{t}] \leq \overline{W} \cap \langle \overline{z} \rangle$ whence $[\overline{W}, \overline{t}] = 1$.

This last case completes the proof of the lemma. □

LEMMA 13.12. *Suppose that Hypothesis* 13.6 *holds, let* $v \in \{a, az\}$ *and suppose that* $t \in C$ *is an involution that is conjugate to* v. *Suppose that* $C \leq H \leq M$ *and that* W *is a nilpotent* $C_H(t)$-*invariant* $2'$-*subgroup of* H. *Then* $[W, t] \leq F(H)$. *In particular if* U_t *is a* t-*minimal subgroup of* G *contained in* H, *then* $U_t \leq F(H)$.

PROOF. Set $\widetilde{H} := H/O(H)$. By Lemma 2.8 we know that
$$O(H) \cap C = O(C_H(z)) = O(C).$$
So as $H = CO(H)$ by Lemma 4.2 (1), we deduce that $\widetilde{H} \simeq C/C \cap O(H) \simeq \overline{C}$. Therefore $F^*(\widetilde{H})$ is isomorphic to $F^*(\overline{C})$ and in particular $\widetilde{t} \in F^*(\widetilde{H})$. This implies that $[\widetilde{W}, \widetilde{t}] \leq F^*(\widetilde{H})$ and hence $[\overline{W}, \overline{t}] \leq F^*(\overline{C})$. As $\overline{W}$ is nilpotent, $C_{\overline{C}}(\overline{t})$-invariant and of odd order, Lemma 13.11 yields that $[\overline{W}, \overline{t}] = 1$. Therefore $[W, t] \leq O(H)$. We deduce from Lemmas 2.9 and 2.1 (2) that $[W, t] = [W, t, t] \leq F(O(H)\langle t \rangle)$. But then $[W, t] \leq F(H)$ because $F(O(H)\langle t \rangle) = F(O(H))$.

The last statement of the lemma follows because $U_t = [U_t, t]$ and U_t is $C_G(t)$-invariant and nilpotent. □

LEMMA 13.13. *Suppose that Hypothesis* 13.6 *holds, that* $C \neq M$ *and that* $p \in \pi(F(M))$ *is such that* $O_p(M)$ *contains a* z-*minimal subgroup* U. *Let* $v \in \{a, az\}$ *and suppose that* $t \in v^C$ *and* $C_G(t) < H_t \, \mathrm{max}\, G$. *Let* U_t *denote a* t-*minimal subgroup of* G *contained in* $F(H_t)$. *If* $U_t \leq C$, *then* M *and* H_t *are both of characteristic* p.

PROOF. Suppose that $U_t \leq C$. Then Lemma 13.12 implies that $U_t \leq C_{F(M)}(z)$. By Lemma 13.7 we may apply Lemma 6.9, so $N_G(U_t) \leq H_t$. Thus $M \looparrowright H_t$. Moreover Lemma 6.7 gives that $U_t \leq C_G(U)$ and hence $U \leq C_G(U_t) \leq H_t$. With the Pushing Down Lemma (3), it follows that $U \leq O_p(H_t)$ and then conversely $H_t \looparrowright M$, by Lemma 6.9. As $C_G(t) \not\leq M$ by Lemma 4.13, we see that H_t and M are distinct. Thus the Infection Theorem (3) forces M and H_t to be of characteristic p as stated. □

LEMMA 13.14. *Suppose that Hypothesis* 13.6 *holds, that H is a maximal subgroup of G and that $C \neq H$. Suppose further that $V \leq H$, that L is a component of H and let $v \in \{a, az\}$. Then the following hold:*

(1) $r_2(E(H)) \leq 2$. *In particular, the 2-rank of L is 1 or 2.*
(2) *If $C_G(v) \neq H_v$ and $U_v \leq H$, then $L \leq H_v$.*
(3) *If $C_G(V) \leq H$, then $L \leq H_v$ or $F^*(\overline{C})$ is as on List IV and $\overline{L}$ coincides with the simple component of $\overline{C}$. Then in particular $L/O(L)$ is isomorphic to the simple component of $\overline{C}$ and $a \in L$.*

PROOF. As $z \in H < G$, Lemma 4.2 (5) yields that $[E(H), z] = 1$. At the same time, the hypothesis that $C \neq H$ implies that $z \notin Z(H)$ and hence $z \notin E(H)$ by Lemma 4.2 (6). Then Theorem **D** gives that $r_2(E(H)) \leq 2$ as stated in (1).

As $E(H) \leq C$, Lemma 13.1 (3) implies that $\overline{E(H)} \leq (O^{2'}(\overline{C}))^\infty = E(\overline{C})$. Assume that $h \in H$ is such that $L^h \neq L$ and set $L_1 := LL^h$. Then $\overline{L_1}$ is a central product of two isomorphic components that is contained in $E(\overline{C})$, so inspection of Lists I-IV for such subgroups yields that $r_2(L) = 1$ and $z \in L$. This implies that $z \in Z(H)$ by Lemma 4.2 (6), contrary to the hypothesis that $C \neq H$. Therefore $L \trianglelefteq H$.

Before we turn to (2), we go through the lists of the F*-Structure Theorem. We bear in mind that in (2) and (3), the subgroup $\overline{L}$ is $C_{\overline{C}}(\overline{V})$-invariant by the previous paragraph.

Lists I and III: As $z \notin L$, the only possible case is that $\overline{L}$ is diagonal in the product of two components of $\overline{C}$. Then there exists an odd number q_0 such that $\overline{L} \simeq PSL_2(q_0)$ and, as $\overline{v} \in F^*(\overline{C})$, either $\overline{v} \in \overline{L}$ or $\overline{v}$ induces an involutory automorphism in $PGL_2(q_0)$ on $\overline{L}$. In both cases $C_{\overline{L}}(\overline{v})$ is a dihedral group.

List II: Here $C_{E(\overline{C})}(\overline{V})$ is of index 2 in a maximal subgroup of $E(\overline{C})$ and therefore $\overline{L}$ is contained in this maximal subgroup. But then $\overline{L} \leq C_{E(\overline{C})}(\overline{V})$ because L is quasi-simple and hence $O^2(L) = L$.

List IV: Let $\overline{E} \leq \overline{C}$ denote the simple component. Then $\overline{v}$ or $\overline{vz}$ is contained in $\overline{E}$ and $\overline{L} \leq \overline{E}$ because $\overline{L}$ is $C_{\overline{C}}(\overline{V})$-invariant. Hence there exists an odd number q_0 such that $\overline{L} \simeq PSL_2(q_0)$ and either $\overline{v} \in \overline{L}$ or $\overline{v}$ induces an involutory automorphism in $PGL_2(q_0)$ on $\overline{L}$. In both cases $C_{\overline{L}}(\overline{v})$ is a dihedral group.

Now suppose that the hypothesis from (2) is satisfied. If L centralises v or U_v, then $L \leq H_v$ by choice of H_v and by Lemma 6.7, which is applicable by Lemma 13.7. Thus we suppose that v and U_v act non-trivially on L. In the cases from List II, we have that $\overline{v}$ centralises $\overline{L}$ and then $[L, v] \leq L \cap O(C) \leq O(L) \leq Z(L)$. Thus v centralises L, contrary to our assumption.

This leaves Lists I, III and IV. If $v \in L$, then $U_v = [U_v, v] \leq L$ because $L \trianglelefteq H$, and then the observations for the lists and the corresponding possibilities for L show that $U_v \leq C_L(v)$. This is impossible. The same argument implies that U_v does not induce inner automorphisms on L. Now we have that $v \notin L$ and inspection of the possibilities above yields that there exists an odd number q_0 such that $L/O(L)$ is isomorphic to $SL_2(q_0)$ or $PSL_2(q_0)$ and such that v induces an involutory automorphism in $GL_2(q_0)$ or $PGL_2(q_0)$. As $\text{Aut}(L)$ is abelian now, we deduce that $[U_v, v] = 1$, which is a contradiction. Thus $L \leq H_v$ and this proves (2).

Suppose that $C_G(V) \leq H$. As $L \trianglelefteq H$, this implies that L is $C_G(V)$-invariant and hence $\overline{L}$ is $C_{\overline{C}}(\overline{v})$-invariant. Thus we know the possibilities for L from the observations before (2). If $\overline{L}$ is diagonally embedded in $F^*(\overline{C})$, then we let $E_1, E_2 \in \mathcal{L}_2(C)$ be such that $\overline{L} \leq \overline{E_1} * \overline{E_2}$ and we deduce that $\overline{L} = [\overline{L}, C_{\overline{E_1}}(\overline{v})] \leq \overline{E_1}$, which is a contradiction. Thus $F^*(\overline{C})$ is as on List II or IV. For the remainder we suppose that $L \not\leq H_v$. Then v does not centralise L and therefore $\overline{v}$ does not centralise $\overline{L}$. This rules out the cases from List II. Now $F^*(\overline{C})$ is as on List IV as stated and there exists an odd number q_0 such that $\overline{L} \simeq PSL_2(q_0)$. In particular we saw that $\overline{L}$ is contained in the simple component $\overline{E}$ of $\overline{C}$ and either $\overline{v} \in \overline{L}$ or $\overline{v}$ induces an involutory automorphism from $PGL_2(q_0)$ on $\overline{L}$. As $C_{\overline{E}}(\overline{V})$ is a maximal subgroup of $\overline{E}$ and leaves $\overline{L}$ invariant, but does not contain $\overline{L}$, it follows that $\overline{L} = \overline{E}$. Therefore $L/O(L) \simeq \overline{E}$ and $a \in L$ as stated, and the proof of (3) is complete. $\square$

13.2. The Symmetric Case

In the remainder of this chapter we prove that maximal subgroups containing the centraliser of an involution a chosen as in Lemma 13.4 either have a central involution or have odd prime characteristic. We begin with the case where both $C_G(a)$ and $C_G(az)$ are not maximal in G.

HYPOTHESIS 13.15.
In addition to Hypothesis 13.6, *suppose that, if* $v \in \{a, az\}$, *then* $C_G(v)$ *is properly contained in* H_v.

With Lemma 13.7, the involutions v appearing in Hypothesis 13.15 satisfy Hypothesis 6.6. It is also worth mentioning that some of the following arguments resemble those in Section 8.

LEMMA 13.16.
Suppose that Hypothesis 13.15 *holds. If* $v \in \{a, az\}$, *then* $O_2(H_v)E(H_v) = 1$.

PROOF. Let $v \in \{a, az\}$. Then by Lemma 13.5 (1), every involution in V is isolated in H_v. Lemma 13.1 (5) and the hypothesis imply that $V \cap Z(H_v) = 1$, so it follows from Lemma 4.2 (6) that $V \cap O_2(H_v)E(H_v) = 1$. Thus no H_v-conjugate of v, of vz or of z can be contained in $O_2(H_v)E(H_v)$. Let $w := vz$.

First we assume that $O_2(H_v) \neq 1$. Then $O_2(H_v)$ has a central involution t and we noticed above that t is not conjugate to z, to v or to w in H_v. As $r_2(G) \leq 3$ by Theorem **D** and as $\langle v, z \rangle \leq C_G(O_2(H_v))$ by Lemma 13.5 (7) whereas none of these involutions lies in $O_2(H_v)$, it follows that t is the unique involution in $O_2(H_v)$. This implies that $t \in Z(H_v)$. In particular, the involution t is central in our Sylow 2-subgroup T_v of H_v which means that T_v contains an elementary abelian subgroup of order 8. Our special choice of a and the 2-structure of C (and hence of H_v) that we can see from the F*-Structure Theorem only leaves very few cases where T_v can contain a central elementary abelian subgroup of order 8. These possibilities are precisely as described in List IV, in the special case where T_v is the direct product of a quaternion group with an elementary abelian group of order 4. But then in C, the involutions t and v or t and w are conjugate. Therefore $C_G(v)$ or $C_G(w)$

must be a maximal subgroup as well, but this contradicts Hypothesis 13.15. Thus $O_2(H_v) = 1$.

Now assume that $E(H_v) \neq 1$ and let L be a component of H_v. As V centralises L and $L \cap V = 1$, Theorem **D** yields that $r_2(L) = 1$. This contradicts the fact that $O_2(H_v) = 1$. □

THEOREM 13.17. *Suppose that Hypothesis* 13.15 *holds. Then* H_a *and* H_{az} *have odd prime characteristic.*

PROOF. Assume otherwise. Then one of H_a, H_{az} does not have odd prime characteristic. For all $v \in \{a, az\}$, we set $F_v := F^*(H_v)$. We know by Lemma 13.16 that $E(H_v)O_2(H_v) = 1$, so in particular $F_v = O(F(H_v))$ and H_v is not of characteristic 2. Hence if H_v is not of odd prime characteristic, then H_v is not of prime characteristic at all, which means that $|\pi_v| \geq 2$. For all $v \in \{a, az\}$ we also set $X_v := [C_{O_{r_v}(H_v)}(w), v]$.

It is used throughout that, by Lemma 13.1 (5), the subgroups M, H_a and H_{az} are pairwise distinct. We also recall that Hypothesis 13.15, which is a special case of Hypothesis 13.6, implies Hypothesis 6.6 by Lemma 13.7.

From now on let $v \in \{a, az\}$ and $w := vz$.

(1) At most one of X_v and X_w is non-trivial. If $X_v \neq 1$, then $r_v \in \pi_w$ and $N_G(X_v) \leq H_w$, hence $H_v \looparrowright H_w$ and $|\pi_w| \geq 2$.

PROOF. As v is isolated in H_w by Lemma 13.5 (1) and as $F_v \cap H_w$ is a nilpotent $C_{H_w}(v)$-invariant subgroup of H_w, we deduce that

$$X_v = [X_v, v] \leq [F_v \cap H_w, v] \leq F_w$$

with the Pushing Down Lemma (2). Hence $X_v \leq O_{r_v}(H_w)$ and $r_v \in \pi_w$. Also, by definition, we have that $[X_v, w] = 1$ and then Lemma 6.7 implies that $X_v \leq C_{F_w}(w) \leq C_G(U_w)$. Therefore X_v is a $U_w\langle w \rangle$-invariant subgroup of F_w.

Suppose that $X_v \neq 1$ and let $N_G(X_v) \leq H \max G$. Then $H_v \looparrowright H$ and $H_w \looparrowright H$. If $H \neq H_w$, then with Lemma 6.10 it follows that H and H_w both have characteristic r_w. Then the Infection Theorem (2), together with the fact that $H_v \looparrowright H$, implies that H_v has characteristic r_w as well. This contradicts the fact that H_a and H_{az} do not both have prime characteristic. Thus $H = H_w$ and $H_v \looparrowright H_w$. In particular the Infection Theorem (2) gives that $|\pi_w| \geq 2$.

For the first statement in (1) assume that $X_w \neq 1$ as well. Then $H_w \looparrowright H_v$ by symmetry and the Infection Theorem (3) yields that $H_v = H_w$. This is impossible because $H_a \neq H_{az}$.

Therefore at most one of X_v and X_w is non-trivial. □

(2) If $X_v = 1$, then $U_v \leq C \cap O_{r_v}(M)$.

PROOF. If $X_v = 1$, then $C_{O_{r_v}(H_v)}(w) \leq C_{O_{r_v}(H_v)}(v)$ and therefore Lemma 2.1 (4) implies that $[O_{r_v}(H_v), v] \leq [O_{r_v}(H_v), w] \cap C$. It follows that

$$U_v = [U_v, v] \leq [O_{r_v}(H_v), v] \leq C.$$

Lemma 13.12 yields that $U_v \leq O_{r_v}(M)$ and hence $U_v \leq C \cap O_{r_v}(M)$. □

(3) $C = M$.

PROOF. Assume otherwise. Then by Hypothesis 13.15 we have a z-minimal subgroup U in M. Applying (2) suppose that $X_v = 1$ and hence $U_v \le C$. Then Lemma 13.13 yields that M and H_v are both of characteristic p. If $X_w = 1$ as well, then it follows from (2) that $U_w \le C$ and Lemma 13.13 forces $\mathrm{char}(H_v) = \mathrm{char}(M) = \mathrm{char}(H_w) = p$, contrary to our assumption. If $X_w \ne 1$, then (1) implies that $|\pi_v| \ge 2$, but we just observed that H_v has characteristic p. So this is impossible and we deduce that $C = M$. □

(4) z centralises $F_\pi(H_v)$. If $X_v = 1$, then z inverts $F_{\pi'}(H_v)$ and $[H_v, z] \le Z(F_v)$.

PROOF. Lemma 4.10 is applicable (by (3)) and gives that $O_\pi(H_v)$ is contained in C. Suppose that $X_v = 1$. Then $U_v \le F(C)$ by (2) and with Lemma 6.9 it follows that $C \looparrowright H_v$. From the Infection Theorem (1) we deduce that $F_{\pi'}(H_v) \cap C = 1$, so z inverts $F_{\pi'}(H_v)$. As $F_v = F^*(H_v)$, this yields that

$$[H_v, z] \le C_{H_v}(F_v) \le Z(F_v).$$
□

(5) If $X_v = 1$, then $F_{\pi'}(H_v)$ is inverted by w and centralised by v.

PROOF. Suppose that $X_v = 1$ and set $Q_v := F_{\pi'}(H_v)$. Then Q_v is abelian because z inverts it by (4). Let $D := C_{Q_v}(w)$.

Suppose that $X_w = 1$ as well. Then (4) implies that $D = [D, z] \le [H_w, z] \le Z(F_w)$ and therefore D is centralised by F_v and by F_w. Moreover D is invariant under $C_C(v) = C_C(w)$, so we have that

$$\langle F_v, F_w, C_C(v) \rangle \le N_G(D).$$

As $z \in Z^*(C_G(v))$, we deduce from (4) that

$$C_G(v) \le C_C(v)[H_v, z] \le C_C(v) Z(F_v)$$

and similarly
$$C_G(w) \le C_C(w) Z(F_w).$$

But this means that $\langle C_G(v), C_G(w) \rangle \le N_G(D)$. With Lemma 13.1 (4) and (1) we deduce first that D is normal in G and then that $D = 1$. Hence if $X_w = 1$, then Q_v is inverted by w and by z and therefore centralised by v. Thus we may suppose that $X_v = 1$ and $X_w \ne 1$. By (1) we have that $H_w \looparrowright H_v$ and that $|\pi_v| \ge 2$. As D is inverted by z and centralised by w, it is inverted by v whence, with the Pushing Down Lemma (2), it follows that

$$D = [D, v] \le [F_v \cap H_w, v] \le F_w.$$

More specifically, we see that $D \le C_{F_w}(w)$ which means that D is a $U_w\langle w \rangle$-invariant subgroup of F_w by Lemma 6.7. Assume that $D \ne 1$ and let $N_G(D) \le H \max G$. Then H_v and H_w infect H. If $H \ne H_w$, then Lemma 6.10 gives that H and H_w have characteristic r_w. Then it follows from the Infection Theorem (2) that $\mathrm{char}(H_v) = r_w$ as well, contrary to $|\pi_v| \ge 2$. Therefore $H = H_w$ which implies that $H_v \looparrowright H_w$. As H_v and

H_w are not of the same prime characteristic, the Infection Theorem (3) leads to a contradiction. Thus $D = 1$ and the proof is complete. □

(6) If $X_v = 1$, then v centralises $F_{r'_v}(H_v)$.

PROOF. Suppose that $X_v = 1$ and assume that $q \in \pi_v$ is such that $q \neq r_v$ and $[O_q(H_v), v] \neq 1$. Then $O_q(H_v)$ contains a v-minimal subgroup U_1 with the same properties as U_v. More specifically, Lemma 6.9 (applied to U_1) yields that $N_G(U_1) \leq H_v$. Now let $Y_v := [C_{O_q(H_v)}(w), v]$.

Assume that $Y_v = 1$. Then we apply (2) to X_v and Y_v and we obtain that $U_v \leq O_{r_v}(C)$ and $U_1 \leq O_q(C)$. Now assume that $U_w \leq O_{r_w}(C)$. Then U_v or U_1 centralises U_w. We may assume that U_w and U_v centralise each other and then $U_w \leq N_G(U_v) \leq H_v$ with Lemma 6.9. Then the Pushing Down Lemma (3) yields that $U_w \leq F_v$. But $N_G(U_w) \leq H_w$, again with Lemma 6.9, so we have that $H_v \looparrowright H_w$. Conversely $U_v \leq N_G(U_w) \leq H_w$ whence $U_v \leq F_w$ and consequently $H_w \looparrowright H_v$. This contradicts the Infection Theorem (3) because H_v and H_w are distinct and not of the same prime characteristic. Consequently $U_w \not\leq O_{r_w}(C)$.

Together with (2) and (4) this implies that $X_w \neq 1$ and that $r_w \notin \pi$. Then it follows from (1) that $H_w \looparrowright H_v$. We push this a little further and look at $U_0 := [C_{U_w}(z), w]$. If $U_0 = 1$, then Lemma 2.1 (4) implies that

$$U_w = [U_w, w] \leq [U_w, z] \cap C_G(v).$$

The Pushing Down Lemma (3) then forces $U_w \leq F_v$ whence $H_v \looparrowright H_w$, by Lemma 6.9. Thus, again, we see that H_v and H_w infect each other, which is a contradiction. We deduce that $U_0 \neq 1$ and we observe that

$$U_0 = [C_{U_w}(z), w] \leq [U_w \cap M, w] \leq F(M),$$

by Lemma 13.12, because $U_w \cap M$ is a nilpotent $C_M(w)$-invariant subgroup of odd order of M. Hence $U_0 \leq C_{F(M)}(z) \leq F(C)$ and it follows that $r_w \in \pi$, contrary to an earlier remark.

We conclude that $Y_v \neq 1$. Then (1), applied to Y_v and X_w, yields that $X_w = 1$ and that $N_G(Y_v) \leq H_w$. Thus $H_v \looparrowright H_w$ and $X_v = X_w = 1$ implies that U_v and U_w are both contained in $F(C)$, by (2). The infection $H_v \looparrowright H_w$ is accomplished by Y_v, more precisely $Y_v \leq O_q(H_v)$ which means that $O_{q'}(H_v) \leq C_G(Y_v) \leq H_w$. But $q \neq r_v$ by our initial assumption, so $U_v \leq O_{q'}(H_v) \leq H_w$ implies first that $U_v \leq F_w$ (with the Pushing Down Lemma (3)) and then that $H_w \looparrowright H_v$ (with Lemma 6.9), which is impossible. This final contradiction shows that v centralises $F_{r'_v}(H_v)$ as stated. □

(7) If $r_v \neq r_w$ and $X_v = 1$, then $X_w \neq 1$ and $r_v \in \pi'_w$, and moreover $F_{r'_w}(H_w)$ is centralised by V.

PROOF. Suppose that $r_v \neq r_w$ and that $X_v = 1$. On the one hand, if $r_v, r_w \in \pi$, then (4) and Lemma 13.12 yield that

$$[U_v, U_w] \leq [O_{r_v}(C), O_{r_w}(C)] = 1.$$

Then it follows with the Pushing Down Lemma (3) and Lemma 6.9 that $U_v \leq F_w$, that $U_w \leq F_v$ and therefore H_v and H_w infect each other. This is impossible by the Infection Theorem (3) and Lemma 13.1 (5). On the other hand (2) implies that $r_v \in \pi$ because $X_v = 1$, so we deduce that $r_w \notin \pi$ and in particular $X_w \neq 1$ by (2).

Now we know from (1) that $H_w \hookrightarrow H_v$, more precisely that $O_{r'_w}(H_w) \leq C_G(X_w) \leq H_v$, and (1) also yields that $|\pi_v| \geq 2$.

First suppose that $|\pi_w| \geq 2$. Then Lemma 6.2 (3) implies that $F_{\pi_w}(H_v) \leq H_w$. Hence if $r_v \in \pi_w$, then it follows that $U_v \leq H_w$. As r_v and r_w are distinct, the Pushing Down Lemma (3) and Lemma 6.9 yield that $U_w \leq C_G(U_v) \leq H_v$ and therefore $U_w \leq F_v$. But then the same argument gives that $H_v \hookrightarrow H_w$. This is a contradiction and hence $r_v \notin \pi_w$ in this case. Next suppose that H_w has characteristic r_w. Then clearly $r_v \notin \pi_w$ because $r_v \neq r_w$.

It is left to prove that $F_{r'_w}(H_w)$ is centralised by V. Let $D := [F_{r'_w}(H_w), z]$. As $1 \neq X_w \leq O_{r_w}(H_w)$ and $N_G(X_w) \leq H_v$ by (1), we have that $D \leq H_v$ and hence

$$D = [D, z] \leq [H_v, z] \leq Z(F_v)$$

by (4). Moreover $D \leq C_G(U_w)$ and $D \leq Z(F_v) \leq C_G(U_v)$. Hence D is a subgroup of $F_v \cap F_w$ that is $U_w\langle w \rangle$-invariant and $U_v\langle v \rangle$-invariant.

If $D \neq 1$, then H_w is not of characteristic r_w and therefore Lemma 6.10 yields that $N_G(D) \leq H_w$, so $H_v \hookrightarrow H_w$. This is a contradiction to the Infection Theorem (3) and Lemma 13.1 (5). Thus $D = 1$. We argue similarly for $D_0 := [F_{r'_w}(H_w), w]$ and recall that $D_0 \leq H_v$. Then

$$D_0 = [D_0, w] \leq [F_w \cap H_v, w] \leq F_v$$

with the Pushing Down Lemma (2) and thus D_0 is a $U_w\langle w \rangle$-invariant subgroup of F_v and of F_w. If $D_0 \neq 1$, then again $|\pi_w| \geq 2$ and Lemma 6.10 and the Infection Theorem (3) give a contradiction. Consequently $D_0 = 1$ and it follows that $F_{r'_w}(H_w)$ is centralised by V. $\square$

(8) If $X_v = 1$, then v inverts $O_{r_v}(H_v)$.

PROOF. Suppose that $X_v = 1$ and in addition that r_v and r_w are distinct. Then (7) yields that $X_w \neq 1$, that $r_v \in \pi'_w$ and that $F_{r'_w}(H_w)$ is centralised by V. In particular (1) gives that $r_w \in \pi_v$ and $H_w \hookrightarrow H_v$. The Infection Theorem (1) implies that $F_{\pi'_w}(H_v) \cap H_w = 1$ and hence that $F_{\pi'_w}(H_v)$ is inverted by w. As $r_v \notin \pi_w$, we conclude that $O_{r_v}(H_v)$ is inverted by w. Moreover (2) and (4) imply that z centralises $O_{r_v}(H_v)$, so v inverts it. Thus we suppose from now on that $r_v = r_w =: r$.

Let $R := O_r(H_w)$. Then $R \neq 1$. Let $D := C_{O_r(H_v)}(v)$ and assume that $D \neq 1$. From (2) we know that $r \in \pi$ and therefore (4) yields that $O_r(H_v) \leq C$. Hence $[D, V] = 1$. As $D \leq C_G(w) \leq H_w$, we can consider the action of $D \times \langle w \rangle$ on R. We know that $[R, w] \neq 1$ because $R = O_r(H_w)$ contains U_w. Thus Thompson's $P \times Q$-Lemma 2.2 forces $R_0 := [C_R(D), w] \neq 1$. From Lemma 6.7 we know that $[D, U_v] = 1$ and hence D is a $U_v\langle v \rangle$-invariant subgroup of F_v. Lemma 6.10 gives that $N_G(D) \leq H_v$ or that H_v has characteristic r. In the second case z centralises $O_r(H_v) = F^*(H_v)$ because $r \in \pi$ by an earlier remark, and this contradicts Lemma 13.1 (5) and Lemma 4.2 (6). This argument also

shows that $|\pi_v| \geq 2$ and $|\pi_w| \geq 2$. We conclude that $N_G(D) \leq H_v$ whence $R_0 \leq H_v$. It follows that
$$R_0 = [R_0, w] \leq [F_w \cap H_v, w] \leq F_v$$
by the Pushing Down Lemma (2). In particular R_0 is $Z(F_v)$-invariant and $C_C(v)$-invariant. Now we apply (4) to see that
$$[C_G(v), z] \leq [H_v, z] \leq Z(F_v)$$
and therefore $C_G(v) \leq Z(H_v)C_C(v)$. It follows that R_0 is $C_G(v)$-invariant.

Let $N_G(R_0) \leq H \max G$. Then $H_v \looparrowright H$ and $C_G(v) \leq H$, in particular v is isolated in H by Lemma 13.5 (1). Therefore $E(H) \leq H_v$ by Lemma 4.2 (5). Hence if $H \neq H_v$, then the Infection Theorem (1) and (4) imply that $F_{\pi_v'}(H)$ is a non-trivial subgroup that is inverted by v. Then it follows from Hypothesis 13.15 that π_v contains a prime p such that $O_p(H_v)$ is inverted by v, and (6) yields that $p = r$. In particular $D = 1$ contrary to our assumption. Therefore $H = H_v$ and $N_G(R_0) \leq H_v$. As $R_0 \leq F_w$, this gives that $H_w \looparrowright H_v$. We recall that $F_w \not\leq C$ and that, therefore, there exists a prime q such that $Q_0 := [O_q(H_w), z] \neq 1$. As $q \neq r$ and $O_{r'}(H_w) \leq C_G(R_0) \leq H_v$, we deduce with (4) that
$$Q_0 = [Q_0, z] \leq [H_v, z] \leq Z(F_v).$$
Hence Q_0 is a non-trivial $U_v\langle v\rangle$-invariant subgroup of F_v and Lemma 6.10 forces $N_G(Q_0) \leq H_v$. But then $U_w \leq C_G(Q_0) \leq H_v$ whence the Pushing Down Lemma (3) and Lemma 6.9 imply that $U_w \leq F_v$ and hence $H_v \looparrowright H_w$. Thus H_v and H_w infect each other, which is impossible.

This last contradiction comes from the assumption that $C_{O_r(H_v)}(v) \neq 1$ and therefore the proof is complete. □

(9) If $X_v = 1$, then $[H_v, V] \leq Z(F_v)$.

PROOF. Suppose that $X_v = 1$. Then $[H_v, z] \leq Z(F_v)$ by (4). Moreover (6) and (8) imply that $[H_v, v] \leq Z(F_v)$, so we have that $[H_v, V] \leq Z(F_v)$. □

(10) If $X_v = 1$, then there exists a prime $p_v \in \pi_v$ such that $O_{p_v}(H_v)$ is centralised by v and inverted by z.

PROOF. Suppose that $X_v = 1$. Then $U_v \leq O_{r_v}(C)$ by (2) and hence $r_v \in \pi$. Lemma 13.1 (5) implies that $z \notin Z(H_v)$, thus we have that $[F_v, z] \neq 1$ by Lemma 4.2 (6). Therefore $\pi_v \cap r_v'$ is not contained in π by (4) and we choose $p_v \in \pi_v \cap r_v' \cap \pi'$. Then $O_{p_v}(H_v)$ is inverted by z by (4) and it is centralised by v by (6). □

(11) If $X_v = 1$ and if $p_v \in \pi_v$ is a prime chosen as in (10), then there exists a V-invariant Sylow p_v-subgroup P of H_v such that $[P, v] = 1$ and $[P, z] = O_{p_v}(H_v)$.

PROOF. Suppose that $X_v = 1$. With Lemma 13.5 (8) let P be a V-invariant Sylow p_v-subgroup of H_v. Then (4) yields that $[P, z] \leq Z(F_v)$. As z inverts $O_{p_v}(H_v)$ and $O_{p_v}(H_v) \leq P$, this means that $[P, z] = O_{p_v}(H_v)$.

We also have that $[P,v] \leq Z(F_v)$ by (9). Therefore, with (6) and Lemma 2.1 (2), we deduce that

$$[P,v] = [P,v,v] \leq P \cap [Z(F_v),v] \leq P \cap O_{r_v}(H_v) = 1.$$

□

(12) There exist a prime p and a V-invariant Sylow p-subgroup P^* of G such that the following hold:
v or w centralises P^* and $[P^*,z]$ equals either $O_p(H_v)$ or $O_p(H_w)$.

PROOF. From (1) we know that $X_v = 1$ or $X_w = 1$, so by symmetry we may suppose that $X_v = 1$. Let $p := p_v$ be as in (10). Let $P \in \mathrm{Syl}_p(H_v, V)$ be as in (11), so that in particular $[P,v] = 1$. We note that $C_G(P) \leq C_G(O_p(H_v)) \leq N_G(O_p(H_v)) = H_v$ because $p \in \pi_v$ and H_v is primitive by Corollary 4.8. Unfortunately Lemma 6.11 is not applicable, but we can argue in a similar way here:

First (11) implies that $[P,z] = O_p(H_v)$. Then, as $z \in N_G(P)$, but $z \notin C_G(P) \trianglelefteq N_G(P)$, it follows that $C_G(P)$ does not contain any conjugate of z. We have that v lies in $C_G(P)$ and that v is isolated in this subgroup because $C_G(P) \leq H_v$. Let $v \in T_0 \in \mathrm{Syl}_2(C_G(P))$. As $C_G(P)$ is z-invariant, we may choose T_0 to be z-invariant by Lemma 3.11. Now T_0 is contained in a Sylow 2-subgroup of H_v and Lemma 3.1 (2) implies that v centralises this Sylow 2-subgroup. Thus $v \in Z(T_0)$. Let $H := N_G(P) \cap N_G(T_0)$. If $H \leq H_v$, then a Frattini argument yields that

$$N_G(P) = C_G(P)H \leq C_G(P)H_v \leq H_v.$$

Thus we suppose that $H \not\leq H_v$, in particular $H \not\leq C_G(v)$. Then v is not the only involution in $Z(T_0)$ because otherwise it is centralised by $N_G(T_0)$. As z centralises T_0, but $z \notin T_0$, it follows that $r(T_0) \leq 2$ because $r_2(G) \leq 3$. So there are exactly three involutions in $Z(T_0)$, which we denote by v, d and vd. We know that $d \in C$ because T_0 is centralised by z. Let $h \in H$ be such that $v^h \neq v$, without loss $v^h = d$. By Lemma 3.1 (9) we may choose h in $C_H(z)$.

Now let $H_d := (H_v)^h$. Then $P = P^h \leq H_d$. As $X_v = 1$, we know that $[H_v, z] \leq Z(F_v)$ by (4), so $[H_d, z] \leq Z(F(H_d))$ by conjugacy. Thus $O_p(H_v) = [P,z] \leq Z(F(H_d))$ and it follows that $O_p(H_v) = O_p(H_d)$. But then $H_v = H_d$ because H_v and H_d are primitive (Corollary 4.8) and therefore $h \in H_v$, which is a contradiction. We conclude that $H \leq H_v$ and thus $N_G(P) \leq H_v$. Therefore $P \in \mathrm{Syl}_p(G)$ and we may choose $P^* = P$.

□

Let p and $P := P^* \in \mathrm{Syl}_p(G, V)$ be as in (12). As $X_v = 1$ or $X_w = 1$ and as this leads to the cases $p \in \pi_v$ or $p \in \pi_w$, we may by symmetry suppose that $X_v = 1$ and hence that $p \in \pi_v$. Then $[P,v] = 1$ and thus $P \leq C_G(v) \leq H_v$. As P is centralised by v, but not by z (from (11)), it follows with Lemma 3.12 that $|C_K(vz)|_p = 1$ and $|C_K(v)|_p \neq 1$. In particular, the involutions v and vz are not conjugate in C, so they are not conjugate in G by Lemma 3.1 (9). Therefore a and az are not conjugate in G. With Lemma 13.5 (2) it follows that $F^*(\overline{C})$ is as on List IV.

13.2. THE SYMMETRIC CASE

Let $E \in \mathcal{L}_2(C)$ be such that $a \in E$. Let $B \leq E$ be elementary abelian of order 4 and such that $a \in B$, and denote the involutions in B by a, d and ad. Then there exists an element $x \in C$ such that $\overline{x}$ has order 3 and such that x permutes a, d and ad in a 3-cycle and then also az, dz and adz in a 3-cycle. Without loss $a^x = d$. Recall that $v \in \{a, az\}$ and define $e := v^x$ and $H_e := H_v^x$. We also recall that z inverts $O_p(H_v)$ and hence $O_p(H_v)$ is abelian. Then the coprime action of e on $O_p(H_v)$ and Lemma 2.1 (4) yield that

$$O_p(H_v) = C_{O_p(H_v)}(e) \times [O_p(H_v), e].$$

Case 1: $D_0 := C_{O_p(H_v)}(e) \neq 1$.

As z inverts $O_p(H_v)$, we know that $D_0 = [D_0, z]$ and that $O_p(H_v)$ is abelian, in particular $[D_0, U_v] = 1$. Consequently D_0 is a $U_v\langle v \rangle$-invariant, non-trivial subgroup of F_v. We recall that $r_v \in \pi$ by (2) and $p \notin \pi$, so in particular $|\pi_v| \geq 2$. Then Lemma 6.10 forces $N_G(D_0) \leq H_v$. As $[H_v, z] \leq Z(F_v)$ by (4), conjugacy yields that

$$D_0 = [D_0, z] \leq [H_e, z] \leq Z(O_p(H_e)).$$

Together this implies that $H_e \looparrowright H_v$. But H_v and H_e are conjugate and they both have no components, by Lemma 13.16, so the Infection Theorem (5) and the hypothesis $|\pi_v| \geq 2$ force $H_v = H_e$. Therefore $x \in H_v$. In particular, the subgroups $C_G(v)$, $C_G(e)$ and $C_G(ev)$ are now all contained in H_v. At least one of the involutions v, e and ev is contained in B, and $F^*(\overline{C})$ is as on List IV, therefore Lemma 13.3 (2) supplies a contradiction.

Case 2: $C_{O_p(H_v)}(e) = 1$.

By choice of p, this means that z and e invert $P_0 := O_p(H_v)$ whence ez centralises it. If we let H_{ez} denote a maximal subgroup of G containing $C_G(ez)$, then $P_0 = [P_0, z] \leq H_{ez}$. Assume that $O(F(C)) \cap H_{ez} = 1$. Then $O(F(C))$ is inverted by ez and therefore this subgroup is abelian. However, we have that $U_v \leq O(F(C))$ by (2) and then $O(F(C)) \leq C_G(U_v) \leq H_v$ with Lemma 6.9. As ez and w are conjugate in C, we also know that w inverts $O(F(C))$ whence it follows that

$$O(F(C)) = [O(F(C)), w] \leq [H_v, w] \leq Z(F_v),$$

by (9). But $N_G(O(F(C))) = C$, so we have that $H_v \looparrowright C$. We recall that $U_v \leq F(C)$ and hence $C \looparrowright H_v$. With the Infection Theorem (3), this implies that $C = H_v$, which is a contradiction.

We deduce that $O(F(C)) \cap H_{ez} \neq 1$ whence Lemma 6.12 is applicable. It gives that $[H_{ez}, z] \leq F(H_{ez})$ and in particular

$$P_0 = [O_p(H_v), z] \leq O_p(H_{ez}).$$

Hence $H_{ez} \looparrowright H_v$ and $O_{p'}(H_{ez}) \leq C_G(P_0) \leq H_v$. We note that H_v is the unique maximal subgroup of G containing $N_G(P_0)$, by Corollary 4.8. But ez is conjugate to $vz = w$, so $O_{r_w}(H_{ez})$ contains an ez-minimal subgroup U_{ez} with $U_{ez} = [U_{ez}, ez]$. As $P_0 \leq C_{F(H_{ez})}(ez)$, it follows with Lemma 6.7 that U_{ez} centralises P_0. Therefore P_0 is a non-trivial $U_{ez}\langle ez \rangle$-invariant subgroup of $F(H_{ez})$.

If $N_G(P_0) \not\leq H_{ez}$, then Lemma 6.10 yields that $\mathrm{char}(H_{ez}) = p$ and that $N_G(P_0)$ is contained in a maximal subgroup distinct from H_{ez} that also has characteristic p. But H_v infects this subgroup, contradicting the Infection Theorem (2) and our hypothesis that $|\pi_v| \geq 2$. Therefore $N_G(P_0) \leq H_{ez}$ and we deduce that $H_v = H_{ez}$. The choice of p implies that v centralises P_0 and therefore $ezv = we$ centralises P_0. With a maximal subgroup H_{we} containing $C_G(we)$ we argue as in the previous paragraph:

If $O(F(C)) \cap H_{we} = 1$, then we inverts $O(F(C))$, so $O(F(C))$ is abelian and hence $O(F(C)) \leq C_G(U_v) \leq H_v$ with Lemma 6.9. As we is conjugate to v or to w in C, one of v or w inverts $O(F(C))$, therefore v and w invert $O(F(C))$. Thus $O(F(C)) \leq Z(F_v)$ by (9) and $H_v \looparrowright C$. We already saw that this is impossible. Hence $O(F(C)) \cap H_{we} \neq 1$ and

$$P_0 = [P_0, z] \leq [H_{we}, z] \leq O_p(H_{we})$$

by Lemma 6.12. So $H_{we} \looparrowright H_v$. Then, in particular, a we-minimal subgroup U_{we} of H_{we} centralises P_0. Thus P_0 is a non-trivial $U_{we}\langle we \rangle$-invariant subgroup of $F(H_{we})$ and we deduce first that $N_G(P_0) \leq H_{we}$ and then that $H_v = H_{we}$. It follows that $C_G(v)$, $C_G(ez)$ and $C_G(we)$ are contained in H_v.

Now we recall that $F^*(\overline{C})$ is as on List IV. Depending on whether $v = a$ or $v = az$, we know that v or ez is contained in B and for all $b \in B$, we just observed that $C_C(b) \leq H_v$. Therefore Lemma 13.3 (2) yields a contradiction.

Hence both cases cannot occur and this concludes the proof of the theorem. □

LEMMA 13.18. *Suppose that Hypothesis* 13.15 *holds. Then H_a and H_{az} have the same odd characteristic.*

PROOF. Set $b := az$ and assume that the statement does not hold. By Theorem 13.17 we may then suppose that there exist distinct odd primes r_a and r_b such that char$(H_a) = r_a$ and char$(H_b) = r_b$. Now the arguments are similar to those at the beginning of the proof of Theorem 13.17.

Let $Y := [C_{O_{r_a}(H_a)}(b), a]$. As a is isolated in H_b by Lemma 13.5 (1), we have that

$$Y = [Y, a] \leq [F(H_a) \cap H_b, a] \leq F(H_b)$$

with the Pushing Down Lemma (2). Therefore $Y \leq O_{r_a}(H_b) = 1$ because $r_a \neq r_b$. It follows that $C_{O_{r_a}(H_a)}(b) \leq C_{O_{r_a}(H_a)}(a)$ and therefore

$$[O_{r_a}(H_a), a] \leq [O_{r_a}(H_a), b] \cap C$$

with Lemma 2.1 (4). We conclude that

$$U_a = [U_a, a] \leq [O_{r_a}(H_a), a] \leq C \leq M.$$

A symmetric argument shows that $U_b \leq M$ and Lemma 13.12 yields that $U_a \leq O_{r_a}(M)$ and $U_b \leq O_{r_b}(M)$. In particular $[U_a, U_b] = 1$ because $r_a \neq r_b$ which implies, with Lemma 6.9 and the Pushing Down Lemma (3), that $U_a \leq F(H_b)$, contrary to $r_a \neq r_b$. □

13.3. The General Case

We begin with two technical lemmas for our further analysis of maximal subgroups containing the centraliser of an involution a that is chosen as in Lemma 13.4 (and similarly the centraliser of az). They play a role in our treatment of the general case and in Chapter 14.

LEMMA 13.19. *Suppose that Hypothesis* 13.6 *holds. Suppose further that* $C \neq M$, *that* $E(M) = 1$ *and that* $C_G(az)$ *is a maximal subgroup of* G. *Then* $[U, a] \neq 1$.

PROOF. Assume otherwise which means that $[U, a] = 1$. First we note that Hypothesis 13.6 implies Hypothesis 6.6 by Lemma 13.7. Now $U \leq H_a$ and the Pushing Down Lemma (3) gives that $U \leq O_p(H_a)$, so that $H_a \hookrightarrow M$ by Lemma 6.9. Our hypothesis that $E(M) = 1$ and Theorem **A** imply that M has characteristic p and hence H_a has characteristic p, by the Infection Theorem (2). In particular a and $b := az$ are not conjugate because $b \in Z(H_b)$ by hypothesis. With Lemma 13.5 (2) this means that one of the cases from List IV holds. Then there exists an odd number $q \geq 11$ such that a lies in a 2-component of C of type $PSL_2(q)$ and in particular there exists an element $c \in C$ of odd order such that a, a^c and $a \cdot a^c = a^{c^2}$ are the three involutions of a fours group. We set $e := a^c$ and $d := ez$, and we let $C_G(e) \leq H_e \max G$ and $C_G(d) =: H_d \max G$.

(*) $[O_p(H_a), a] \leq C$.

PROOF. Let $X := [C_{O_p(H_a)}(b), a]$ and assume that $X \neq 1$. Then the Pushing Down Lemma (2) yields that

$$X = [X, a] \leq [O_p(H_a) \cap H_b, a] \leq O_p(H_b).$$

With Lemma 13.1 (1) we let $N_G(X) \leq H \max G$. Then $H_a \hookrightarrow H$ and $H_b \hookrightarrow H$ and therefore the fact that H_a is of characteristic p and $p \in \pi_b$ implies (using the Infection Theorem (1)) that $F := F_{\pi_b'}(H)$ is inverted by a and b and hence centralised by z. In particular $F \leq C$ and therefore F normalises U. But a centralises U by assumption whence $F = [F, a]$ centralises U as well. If $F \neq 1$, then consequently $U \leq C_G(F) \leq H$ because H is primitive by Corollary 4.8. Then the Pushing Down Lemma (3) gives that $U \leq O_p(H)$ and hence $H \hookrightarrow M$ by Lemma 6.9. The Infection Theorem (2) first implies that H is of characteristic p and then that H_b is of characteristic p as well. But $H_b = C_G(b)$, so this is impossible.

Therefore $F = 1$ and hence $F(H)$ is a π_b-group. We note that X is $C_G(V)$-invariant and hence $C_G(V) \leq H$, which makes Lemma 13.14 (3) applicable. If $E(H) \not\leq H_v$, then this result yields that H has a component L such that $L/O(L)$ is isomorphic to the simple component of $\overline{C}$ and in particular L contains a. We know from Lemma 6.2 (3) and because $\pi(F(H)) \subseteq \pi_b$ that $F(H) \leq H_b = C_G(b)$. As $a \in L$ (still assuming that $L \not\leq H_v$), this means that a and b centralise $F(H)$ and hence z does. Thus $z \in Z(H)$ by Lemma 4.2 (6) and it follows that $C = H$ is a maximal subgroup, contradicting our hypothesis. We deduce that $E(H) \leq H_b$ and the Infection Theorem (4) yields that $H = H_b$. In particular H_a infects H_b. Then the Infection Theorem (1) gives that a inverts $F_{p'}(H_b)$, but this is false because a centralises $O_2(H_b)$ by Lemma 13.5 (7). Hence $X = 1$ and Lemma 2.1 (4) implies that $[O_p(H_a), a] \leq [O_p(H_a), b] \cap C$. □

We recall that H_a has characteristic p and that therefore $C_G(a) < H_a$. Then $O_p(H_a)$ contains an a-minimal subgroup U_a by Hypothesis 13.6 and we let $U_0 := [C_{U_a}(d), a]$, with d being conjugate to b as described in the first paragraph. As $U_0 \le U_a = [U_a, a] \le C$ by (*), we see that $U_0 = [C_{U_a}(e), a]$. This implies that U_0 is $C_G(\langle a, e \rangle)$-invariant and $C_G(\langle a, d \rangle)$-invariant. Lemma 13.12 yields that $U_a \le O_p(C)$ and therefore

$$U_0 \le O(C) \cap H_d \le O(C_C(d)) \le O(C_G(d)) = O(H_d)$$

by Lemma 2.8 and because d and b are conjugate. Hence $U_0 = [U_0, a] \le [O(H_d), a]$. We apply Lemma 2.9 to $H := O(H_d)\langle a, d \rangle$ and see that $U_0 = [U_0, a] \le O_p(H) = O_p(H_d)$. In a similar way, we deduce that $U_0 \le O(H_e)$ and then $U_0 \le O_p(H_e)$. As e is conjugate to a, we may suppose that H_e is conjugate to H_a and hence has characteristic p. In particular $C_G(e) < H_e$. Then $O_p(H_e)$ contains an e-minimal subgroup U_e and $U_0 \le C_{O_p(H_e)}(e) \le C_G(U_e)$ by Lemma 6.7. Moreover U_0 is $\langle e \rangle$-invariant. Hence if $U_0 \ne 1$, then Lemma 6.10, applied to H_e and U_0, gives that $N_G(U_0)$ is contained in a maximal subgroup of G of characteristic p (which might coincide with H_e, but not necessarily). But H_d infects this subgroup, $O_p(H_d) \ne 1$ and H_d is not of characteristic p, so the Infection Theorem (2) yields a contradiction.

Thus $U_0 = 1$. Then $C_{U_a}(d) \le C_{U_a}(a)$ and Lemma 2.1 (4) implies that $U_a = [U_a, a] \le [U_a, d] \cap C_G(ad)$. We recall that $U_a \le O(C)$ and hence

$$U_a \le O(C) \cap C_G(ad) \le O(C_C(ad)) \le O(C_G(ad))$$

by Lemma 2.8. It follows that $U_a \le O_p(C_G(ad))$, again with Lemma 2.9. As ad is conjugate to b, we know that $C_G(ad)$ is a maximal subgroup of G and Lemma 6.9 yields that $C_G(ad)$ infects H_a. This is impossible by the Infection Theorem (2) because H_a has characteristic p, but $C_G(ad)$ is conjugate to $C_G(b)$ and hence does not have characteristic p. $\square$

LEMMA 13.20. *Suppose that Hypothesis 10.1 holds and that $a \in C$ is an involution chosen as in Lemma 13.4. Suppose that C is not a maximal subgroup of G and let $v \in C$ be an involution that is conjugate to a or to az. Then $C_G(v)$ is a maximal subgroup of G or $C_G(v)$ is contained in a maximal subgroup H_v such that H_v has odd prime characteristic.*

PROOF. As v is conjugate to a or to az, we may without loss suppose that $v \in \{a, az\}$ and we let $w := vz$. Then we suppose that Hypothesis 13.6 holds with its notation. Let $p \in \pi$ be such that $U \le O_p(M)$.

If $C_G(v) \ne H_v$ and v and w are conjugate, then Hypothesis 13.15 holds whence Theorem 13.17 is applicable. Thus in this case H_v is of odd prime characteristic as stated. Hence from now on we suppose that v and w are not conjugate, which by Lemma 13.5 (2) means that $F^*(\overline{C})$ is as on List IV. By choice of a, there exists an odd number $q \ge 11$ such that either v or w is contained in some $E \in \mathcal{L}_2(C)$ with $\overline{E} \simeq PSL_2(q)$.

Let us suppose that $C_G(v) \ne H_v$ and, by way of contradiction, that $|\pi_v| \ge 2$. Let $r := r_v$, so that $U_v \le O_r(H_v)$. We recall that Theorem 13.17 implies that $C_G(w)$ is a maximal subgroup. Let $U_0 := [C_U(v), z]$ and $b := az$.

(1) $1 \ne U_0 \le O_p(H_v)$.

13.3. THE GENERAL CASE

PROOF. First suppose that $v = a$. Then v lies in an elementary abelian subgroup B of C such that some element of C induces a 3-cycle on the set of involutions of B. Then the coprime action of B on U, together with Lemma 2.1 (4), yields that there exists an involution $d \in B$ such that $[C_U(d), z] \neq 1$. Hence $U_0 \neq 1$ in this case.

Next suppose that $v = b$. Then $w = a$, hence $C_G(a)$ is a maximal subgroup. If $U_0 = 1$, then we first assume that M has characteristic p. As $[C_U(b), z] = 1$, we have that $C_U(b) \leq C_U(z)$ whence $U = [U, z] \leq [U, b] \cap C_G(a)$, by Lemma 2.1 (4). Then the Pushing Down Lemma (3) forces $U \leq O_p(H_a)$ and with Lemma 6.9 this implies that $H_a \looparrowright M$. But this is impossible by the Infection Theorem (2) because $2 \in \pi(F(H_a))$ and therefore H_a is not of characteristic p. Therefore M is not of characteristic p. Then M has a simple component by Lemma 13.9, and in fact E is this component. Then $a \in E$ by hypothesis and $[U, a] \leq [O_p(M), E] = 1$. Thus the Pushing Down Lemma (3) and Lemma 6.9 give that $U \leq O_p(H_a)$ and then $H_a \looparrowright M$. We turn to H_b and let $X := [C_{O_r(H_b)}(z), b]$. Then X is a nilpotent $C_C(b)$-invariant $2'$-subgroup of C and Lemma 13.12 yields that $X = [X, b] \leq F(C)$. As $E \leq E(C)$, we see that $[X, E] = 1$. In particular X centralises a and z, so we deduce that $X = [X, b] = 1$. Therefore $C_{O_r(H_b)}(z) \leq C_{O_r(H_b)}(b)$ which with Lemma 2.1 (4) implies that

$$U_b \leq [O_r(H_b), b] \leq [O_r(H_b), z] \cap C_G(a).$$

With the Pushing Down Lemma (3) and Lemma 6.9 it follows first that $U_b \leq O_r(H_a)$ and then that $H_a \looparrowright H_b$. Moreover $r \in \pi_a$. We recall that $F^*(\overline{C})$ is as on List IV and therefore Lemma 13.8 forces $E(H_b) = 1$. As H_a, H_b are neither equal nor of the same prime characteristic, the Infection Theorem (4) yields that $F := F_{\pi'_a}(H_b) \neq 1$.

We can say more if we recall that $U \leq O_p(H_a)$: if $r \neq p$, then $[U, U_b] \leq [O_p(H_a), O_r(H_a)] = 1$ whence $U \leq N_G(U_b) \leq H_b$, then $U \leq O_p(H_b)$ by the Pushing Down Lemma (3) and finally $H_b \looparrowright M$ by Lemma 6.9. Conversely $U_b \leq C_G(U) \leq M$ by Lemma 6.9, then $U_b \leq F(M)$ by Lemma 13.12 and therefore $M \looparrowright H_b$ by Lemma 6.9. Thus $M \looparrowright H_b \looparrowright M$, which contradicts the Infection Theorem (3). Thus $r = p$ and moreover, as we can argue in the same way if $F_{p'}(H_b)$ contains a b-minimal subgroup, we also see that b centralises $F_{p'}(H_b)$.

Now we recall that $a \in E$ and that a is contained in a fours group in E where all involutions are conjugate in E. Let $e \in E$ be an involution such that $\langle a, e \rangle$ is such a fours group. Then a, e and ae are conjugate in C and therefore b, ez and be are conjugate in C as well. We also recall that $E(H_b) = 1$ and that $H_a \looparrowright H_b$. Therefore F is inverted by a (by the Infection Theorem (1)) and $F_{\pi_a}(H_b)$ is centralised by a by Lemma 6.2 (3), because $C_G(a) = H_a$ and $|\pi_a| \geq 2$. In particular

$$[H_b, a] \leq C_{H_b}(F(H_b)) = C_{H_b}(F^*(H_b)) = Z(F(H_b)).$$

Let $H_e := C_G(e)$ and $H_{ae} := C_G(ae)$, and let $C_G(ez) \leq H_{ez} \max G$ and $C_G(be) \leq H_{be} \max G$. Then conjugacy yields that

$$[H_{ez}, e] \leq Z(F(H_{ez})) \text{ and } [H_{be}, ae] \leq Z(F(H_{be})).$$

We have seen earlier that $F \leq F_{p'}(H_b)$ is centralised by b and inverted by a, so F is inverted by z. Now we consider the action of $\langle z, e \rangle$ on $F = [F, z]$ and apply Lemma 2.1 (4). It gives that

$$F = \langle [C_F(z), z], [C_F(e), z], [C_F(ez), z] \rangle = \langle [C_F(e), z], [C_F(ez), z] \rangle.$$

Assume that $Y := [C_F(ez), z] \neq 1$. Then we first note that

$$Y = [Y, z] = [Y, e] \leq [H_{ez}, e] \leq Z(F(H_{ez})).$$

Hence if we let U_{ez} denote an ez-minimal subgroup of $O_p(H_{ez})$, then Y is $U_{ez}\langle ez \rangle$-invariant and we also recall that H_b, and hence H_{ez}, is not of prime characteristic. Therefore Lemma 6.10 is applicable and gives that $N_G(Y) \leq H_{ez}$. In particular $H_b \looparrowright H_{ez}$ and this contradicts Lemma 13.10. Hence $Y = 1$ which implies that $F = [C_F(e), z]$ is centralised by e. Then F is centralised by b and e and inverted by z, hence centralised by be and therefore inverted by ae. We conclude that

$$F = [F, ae] \leq [H_{be}, ae] \leq Z(F(H_{be})).$$

As H_b is primitive by Corollary 4.8, we know that $N_G(F) = H_b$ and therefore $H_{be} \looparrowright H_b$. But this is impossible by Lemma 13.10. This last contradiction comes from the fact that $F \neq 1$, as we established earlier. That in turn was a consequence of the assumption that $U_0 = 1$. So we proved that $U_0 \neq 1$ as stated.

For the second assertion in (1) we apply the Pushing Down Lemma (2). It follows that $U_0 = [U_0, z] \leq [F(M) \cap H_v, z] \leq O_p(H_v)$. □

(2) $N_G(U_0) \leq H_v$.

PROOF. With Lemma 13.1 (1) let H be a maximal subgroup of G containing $N_G(U_0)$.

(2.1) $U_v \leq H$, furthermore $F(H)$ is a π_v-group and lies in H_v:

We have that $M \looparrowright H$ and also, by (1), that $H_v \looparrowright H$. As $U_0 \leq F(H_v)$ by (1), Lemma 6.7 yields that

$$U_0 \leq C_{F(H_v)}(v) \leq C_G(U_v)$$

and therefore $U_v \leq H$. With the Infection Theorem (1) it follows that $F_{\pi_v'}(H) \cap H_v = 1$ and hence that v inverts $F := F_{\pi_v'}(H)$. But then $U_v = [U_v, v]$ centralises F which means, by Lemma 6.9, that $F \leq H_v$ and thus $F = 1$. Consequently $F(H)$ is a π_v-group that is contained in H_v by Lemma 6.2 (3), because $|\pi_v| \geq 2$ by assumption.

(2.2) $E(H) \leq H_v$:

We know from (2.1) that $U_v \leq H$. If $C = H$, then in particular $U_0 \leq C$ and hence $U_0 = [U_0, z] = 1$ by Lemma 2.1 (2). This contradicts (1). Therefore Lemma 13.14 (2) is applicable and yields that $E(H) \leq H_v$.

(2.3) $H = H_v$:

This follows from (2.1) and (2.2) together with the Infection Theorem (4), because $|\pi_v| \geq 2$ by assumption.

Now the proof of (2) is complete. □

(3) Let $X := [C_{O_r(H_v)}(w), v]$. Then $1 \neq X \leq O_r(H_w)$ and $N_G(X) \leq H_v$.

PROOF. First assume that $X = 1$. Then
$$C_{O_r(H_v)}(w) \leq C_{O_r(H_v)}(v) \cap C$$
and hence $[O_r(H_v), v] \leq [O_r(H_v), w] \cap C$ by Lemma 2.1 (4). In particular $U_v \leq C$ whence Lemma 13.13, together with (1), implies that M and H_v both have characteristic p. This contradicts our assumption that $|\pi_v| \geq 2$, hence $X \neq 1$.

With the Pushing Down Lemma (2) we have that $X \leq O_r(H_w)$. Applying Lemma 13.1 (1) let $N_G(X) \leq H_1 \max G$. Then $H_v \hookrightarrow H_1$ and $H_w \hookrightarrow H_1$. Assume that $r = p$. We know that $\pi(F(H_w))$ contains 2 and $r (= p)$ and therefore two distinct primes, and then Lemma 6.2 (3) yields that $O_p(H_1) \leq H_w = C_G(w)$. In particular U_0 is centralised by w, but also by v and thus by z. This is impossible by (1) and hence $r \neq p$. Now we see that $U_0 \leq C_G(X) \leq H_1$ and therefore $U_0 \leq O_p(H_1)$ by the Pushing Down Lemma (2). It follows that $H_1 \hookrightarrow H_v$ by (2) and then that $H_1 = H_v$ by the Infection Theorem (3). □

With (3) we have that $H_w \hookrightarrow H_v$, in particular $O_{r'}(H_w) \leq H_v$. As r is odd and $w \in Z(H_w)$, we see that $|\pi_w| \geq 2$. Let $F := F_{\pi'_w}(H_v)$. Then $F \neq 1$ by the Infection Theorem (4), because $E(H_v) = 1$ by Lemma 13.16 and because $H_v \neq H_w$ by Lemma 13.1 (5). Hence the Infection Theorem (1) yields that w inverts F. Since $|\pi_w| \geq 2$, we see that $O_p(H_v)$ is either inverted or centralised by w (Infection Theorem (1) and 6.2 (3)). As $U_0 \leq O_p(H_v)$ and U_0 is not centralised by w, it follows that w inverts $O_p(H_v)$. Thus $p \in \pi'_w$ whence, with the Pushing Down Lemma (2), we deduce that $[C_U(w), z] \leq [U \cap H_w, z] \leq O_p(H_w) = 1$.

Then $U = [U, z] \leq [U, w] \cap C_G(v)$ by Lemma 2.1 (4) and consequently $U \leq O_p(H_v)$ by the Pushing Down Lemma (3). Then $H_v \hookrightarrow M$ by Lemma 6.9. But conversely $M \hookrightarrow H_v$ by (2), so with the Infection Theorem (3) this contradicts Lemma 13.1 (5). □

LEMMA 13.21. *Suppose that Hypothesis 10.1 holds and that $a \in C$ is an involution chosen as in Lemma 13.4. Let $v \in C$ be an involution that is conjugate to a or to az and suppose that C is a maximal subgroup of G. Let H_v be a maximal subgroup of G containing $C_G(v)$ and suppose that $O(F(C)) \cap H_v = 1$. Then $C_G(v) = H_v$ or H_v is of odd prime characteristic.*

PROOF. Assume that $C_G(v) \neq H_v$ and that H_v is not of odd prime characteristic. If $C_G(vz)$ is properly contained in a maximal subgroup, then Hypothesis 13.15 is satisfied, so that Theorem 13.17 is applicable and yields the result. Thus we suppose that $C_G(vz)$ is a maximal subgroup of G and that Hypothesis 13.6 holds with all its notation. We have that v and vz are not conjugate which means, by Lemma 13.5 (2), that $F^*(\overline{C})$ is as on List IV. Then we may suppose that v or vz coincides with a and hence lies in a 2-component of C with simple image in $\overline{C}$. Let $w := vz$. From the shape of $O_{2',F^*}(C)$, there exists an elementary abelian subgroup of C containing a and such that some element $x \in C$ induces a 3-cycle on the set of involutions of this subgroup. Then x moves az in a 3-cycle with two other involutions, but these involutions do not belong to an elementary abelian group of order 4. In the following, we therefore argue that v, $d := v^x$ and d^x are conjugate, but we might need to distinguish the cases where $v = a$ or $v = az$. Our hypothesis gives that v inverts $O(F(C))$ and hence that w inverts $O(F(C))$. Now d, d^x and then also dz and $d^x z$ invert $O(F(C))$ as well. But then $dv = dzw$ centralises and inverts $O(F(C))$, so it follows that $O(F(C)) = 1$. Then $F(O(C)) = 1$ which forces $O(C) = 1$. Looking at List IV in the F*-Structure Theorem, we let E denote the simple component of C. Its type yields that, if $s, t \in E$ are two distinct involutions, then $E = \langle C_E(s), C_E(t)\rangle$ (recall that $E \simeq PSL_2(q)$ with $q \geq 11$).

Let $Y := [C_{O_{r_v}(H_v)}(z), v]$. This is a $C_C(v)$-invariant $2'$-subgroup of $F(H_v)$ and Lemma 13.12 implies that $Y = [Y, v] = 1$ and consequently $C_{O_{r_v}(H_v)}(z) \leq C_{O_{r_v}(H_v)}(v)$. Then Lemma 2.1 (4) gives that $U_v = [U_v, v] \leq [U_v, z] \cap C_G(w)$.

With the Pushing Down Lemma (3) and Lemma 6.9, we deduce that $U_v \leq F(H_w)$ and therefore $H_w \looparrowright H_v$. Hence $F_{\pi'_w}(H_v)$ is inverted by w by the Infection Theorem (1). We recall that $F^*(\overline{C})$ is as on List IV and that, therefore, Lemma 13.8 yields that $E(H_v) = 1$. Moreover Lemma 6.2 (3) implies that $F_{\pi_w}(H_v)$ is centralised by w because $|\pi_w| \geq 2$. Therefore

$$[H_v, w] \leq C_{H_v}(F(H_v)) = Z(F(H_v)).$$

Let $H_d := H_v^x$, which is a maximal subgroup of G containing $C_G(d)$. If we let U_d denote a d-minimal subgroup of G in $F(H_d)$, then conjugacy yields that $U_d \leq C_G(dz)$ and

$$[H_d, dz] \leq C_{H_d}(F(H_d)) = Z(F(H_d)).$$

Case 1: $[C_{U_v}(d), v] \neq 1$.

As dz and w centralise U_v, we see that

$$U_0 := [C_{U_v}(d), v] = [C_{U_v}(d), z] = [C_{U_v}(d), bz] \leq [H_d, dz] \leq Z(F(H_d)).$$

We know that U_d lies in $O_r(H_d)$ and that $[U_0, U_d] = 1$ by Lemma 6.7, hence U_0 is $U_d \langle d \rangle$-invariant. Applying Lemma 6.10 to H_d and using that $|\pi(F(H_d))| \geq 2$ (because d is conjugate to v), we see that $N_G(U_0) \leq H_d$. But $U_0 \leq F(H_v)$, therefore $H_v \looparrowright H_d$. We have that $E(H_v) = 1 = E(H_d)$ by Lemma 13.8 and $\pi(F(H_d)) = \pi_v$ (with at least two distinct primes!), so that the Infection Theorem (5) gives that $H_v = H_d$. We noted above that the simple component E of C is generated by two different involution centralisers. We also know that v and d or w and dz are contained in E and therefore

$$E = \langle C_E(v), C_E(d)\rangle = \langle C_E(w), C_E(dz)\rangle.$$

As $C_G(v)$ and $C_G(d)$ are both contained in H_v, we deduce that $E \leq H_v$. Then also $E \leq H_w$ and in particular x can be chosen to lie in $H_v \cap H_w$. This is impossible by Lemma 13.3.

Case 2: $[C_{U_v}(d), v] = 1$.

We apply Lemma 2.1 (4) to the coprime action of $\langle d, z\rangle$ on U_v and obtain that

$$U_v = \langle [C_{U_v}(z), v], [C_{U_v}(d), v], [C_{U_v}(dz), v]\rangle.$$

We also recall that $C_{O_{r_v}(H_v)}(z) \leq C_{O_{r_v}(H_v)}(v)$, hence $[C_{U_v}(z), v] = 1$ and therefore the present case implies that $U_v = [C_{U_v}(dz), v]$. Now we consider the action of $\langle z, dv\rangle$ on U_v. Lemma 2.1 (4) yields that

$$U_v = \langle [C_{U_v}(z), v], [C_{U_v}(dv), v], [C_{U_v}(dw), v]\rangle = \langle [C_{U_v}(dv), v], [C_{U_v}(dw), v]\rangle,$$

again because $[C_{U_v}(z), v] = 1$. Let $C_G(dv) \leq H_{dv} := H_d^x$. If $v = a$, then $d^x = dv$ and we let $U_1 := [C_{U_v}(dv), v]$. We recall that dz and w centralise U_v and we see, with the conjugacy of H_v and H_{dv}, that

$$U_1 = [U_1, v] = [U_1, z] = [U_1, dw] \leq [H_{dv}, dw] \leq Z(F(H_{dv})).$$

If $U_1 \neq 1$, then with U_{dv} denoting a dv-minimal subgroup of G contained in $F(H_{dv})$, we deduce that U_1 is a $U_{dv}\langle dv \rangle$-invariant non-trivial subgroup of $F(H_{dv})$. Lemma 6.10 implies that $N_G(U_1) \leq H_{dv}$ and therefore $H_v \looparrowright H_{dv}$. This contradicts Lemma 13.10. In this case we obtain that $U_1 = 1$ and hence that

$$U_v = [C_{U_v}(dw), v].$$

In particular U_v is now centralised by w, by dz and by dw. This implies that U_v is centralised by d, then by z and then by $wz = v$. But this is impossible.

If $v = az$, then $C_G(dv)$ is conjugate to $C_G(w)$ and hence a maximal subgroup. We go back and recall that $U_v \leq H_w$ and then $H_w \looparrowright H_v$. We know therefore, with the Infection Theorem (1) and Lemma 6.2 (3), that $F_{\pi'_w}(H_v)$ is inverted by w and that $F_{\pi_w}(H_v)$ is centralised by w. As $E(H_v) = 1$ by Lemma 13.8, this means that $[H_v, w] \leq Z(F(H_v))$. By conjugacy we have that $[H_{dv}, dw] \leq Z(F(H_{dv}))$. We recall that U_v is centralised by w and by dz and therefore by $dzw = dv$. Therefore

$$U_v = [U_v, v] = [U_v, d] = [U_v, dw] \leq [H_{dv}, dw] \leq Z(F(H_{dv}))$$

whence, with Lemma 6.9, it follows that $H_{dv} \looparrowright H_v$. This means, again with the Infection Theorem (1) and Lemma 6.2 (3), that $F_{\pi'_w}(H_v)$ is inverted by dz (and w) and therefore centralised by dv and that $F_{\pi_w}(H_v)$ is centralised by dz (and w) and therefore by dv. This implies that dv centralises $F(H_v) = F^*(H_v)$ (with Lemma 13.8) and is therefore contained in $O_2(H_v)$. Lemma 13.1 (5) and Lemma 13.5 (1) imply that $V \cap O_2(H_v) = 1$ and therefore Theorem **D** forces dv to be the unique involution in $O_2(H_v)$. In particular $dv \in Z(H_v)$ and therefore $H_v = C_G(dv)$. We recall that, if $s, t \in E$ are two distinct involutions, then $E = \langle C_E(s), C_E(t) \rangle$. Applying this to the involutions $w, dv \in E$, we deduce that

$$E = \langle C_E(w), C_E(dv) \rangle = \langle C_E(v), C_E(dv) \rangle \leq H_v.$$

In particular E is centralised by dv, which is impossible.

We arrived at a contradiction in both cases and therefore the proof is finished. □

LEMMA 13.22. *Suppose that Hypothesis 10.1 holds, that $a \in C$ is an involution chosen as in Lemma 13.4 and that C is a maximal subgroup of G. Let $v \in C$ be an involution that is conjugate to a or to az and let H_v be a maximal subgroup of G containing $C_G(v)$. Suppose that $O(F(C)) \cap H_v \neq 1$. Then $C_G(v) = H_v$ or H_v is of odd prime characteristic.*

PROOF. We begin as for the previous lemma by supposing that $C_G(v) \neq H_v$ and that $C_G(vz)$ is a maximal subgroup of G. In particular v and vz are not conjugate, so by Lemma 13.5 (2) one of the cases from List IV holds. We may suppose that $v \in \{a, az\}$ and hence that Hypothesis 13.6 holds with all its notation.

In particular let $V := \langle z, v \rangle$ and $w := vz$. We assume that H_v is not of odd prime characteristic. The shape of $O_{2',F^*}(C)$ implies that there exists an elementary abelian subgroup of C containing a and such that some element $x \in C$ induces a 3-cycle on the set of involutions of this subgroup. In the following, we therefore argue that v, $d := v^x$ and d^x are conjugate. We let $C_G(d) \le H_d := H_v^x$ and similarly $C_G(vd) \le H_{vd}$, where H_{vd} is conjugate to H_v or to H_w, depending on whether $v = a$ or $v = az$. Moreover let $r := r_v$, so that $U_v \le O_r(H_v)$.

By Lemma 6.12 we have that $[H_v, z] \le F_{\pi'}(H_v)$ and similarly $[H_d, z] \le F_{\pi'}(H_d)$. Assume that v or w inverts $O(F(C))$. Then a inverts it and therefore, with our notation from above, we see that a^x and a^{x^2} also invert $O(F(C))$. But $a^{x^2} = aa^x$ centralises $O(F(C))$ and thus we must have that $O(F(C)) = 1$, which contradicts our hypothesis. In particular w does not invert $O(F(C))$ whence $O(F(C)) \cap H_w \neq 1$ as well. Another application of Lemma 6.12 yields that $[H_w, z] \le F_{\pi'}(H_w)$. We use this relation for all involutions conjugate to a or az.

(1) $E(H_v) = 1$.

PROOF. We know that $F^*(\overline{C})$ is as on List IV and therefore Lemma 13.8 yields the result. □

(2) $U_v \le C$.

PROOF. Assume otherwise. Then $r \notin \pi$ by Lemma 4.10 and Lemma 13.12 implies that $[C_{U_v}(z), v] = 1$. Thus

$$U_v = [U_v, v] \le [U_v, z] \cap C_G(w)$$

by Lemma 2.1 (4). Then the Pushing Down Lemma (3) implies that $U_v \le O_r(H_w)$ and with Lemma 6.9 it follows that $H_w \looparrowright H_v$.

Let $U_1 := [C_{U_v}(d), z]$ and assume that $U_1 \neq 1$. We have that

$$U_1 = [U_1, z] \le [H_d, z] \le F(H_d)$$

and therefore $U_1 \le C_{F(H_d)}(d)$. If we set $U_d := U_v^x$, then U_d is a d-minimal subgroup and we see with Lemma 6.7 that U_1 is centralised by U_d. Hence U_1 is $U_d \langle d \rangle$-invariant and $N_G(U_1) \le H_d$ by Lemma 6.10 (and because $\pi(F(H_d)) = \pi_v$ contains at least two distinct primes). But then $H_v \looparrowright H_d$ and this contradicts Lemma 13.10. Therefore $U_1 = 1$.

The coprime action of $\langle v, d \rangle$ on U_v and Lemma 2.1 (4) then yield that

$$U_v = [U_v, z] = \langle [C_{U_v}(v), z], [C_{U_v}(d), z], [C_{U_v}(vd), z] \rangle = [C_{U_v}(vd), z].$$

This means that $U_v \le C_G(vd)$ and hence (with $C_G(vd) \le H_{vd} \max G$) it follows that

$$U_v = [U_v, z] \le [H_{vd}, z] \le C_{F(H_{vd})}(dv).$$

If $v = a$, then vd is conjugate to v. Lemma 6.9 yields that $H_{vd} \looparrowright H_v$, which is a contradiction to Lemma 13.10. We conclude that if $U_v \not\le C$, then $v = az$ and vd is not conjugate to v, but to w. Moreover we recall that U_v is centralised by w, by vd and (hence) by dz. In particular $H_w \looparrowright H_v$ and we recall that $U_v \le F(H_{vd})$. Similarly, with $C_G(dz) \le H_{dz} \max G$, we have that $U_v = [U_v, z] \le F(H_{dz})$. Thus it follows that H_{vd} and H_{dz} infect H_v as well. Let $F := F_{\pi'_w}(H_v)$. Then $F \neq 1$ by (1), the Infection Theorem (4) and Lemma 13.1 (5). The Infection Theorem (1) implies that F is inverted by w, by vd and by dz, because $\pi_w = \pi(F(H_{dz})) =$

$\pi(F(H_{vd}))$. But if w and vd invert F, then $dz = wvd$ centralises it and therefore $F = 1$. This is a contradiction.

We conclude that $U_v \leq C$. □

(3) C infects H_v, H_d and H_d^x.

PROOF. Together with Lemma 13.12, statement (2) yields that $U_v \leq O_r(C)$. In particular $r \in \pi$ and, as $N_G(U_v) \leq H_v$ with Lemma 6.9, we see that $C \hookrightarrow H_v$. By conjugacy, we also have that C infects H_d and H_d^x. □

(4) $[H_v, z] \leq Z(F(H_v))$.

PROOF. From (3) and the Infection Theorem (1) it follows that z inverts $F_{\pi'}(H_v)$. Moreover z centralises $F_\pi(H_v)$ by Lemma 4.10. As $E(H_v) = 1$ by (1), we deduce that $[H_v, z] \leq C_{H_v}(F^*(H_v)) = Z(F(H_v))$. □

(5) $[F_{\pi'}(H_v), v] = 1$.

PROOF. Assume that this is false and choose $p \in \pi_v \cap \pi'$ such that $[O_p(H_v), v] \neq 1$. Then $P_0 := O_p(H_v)$ is inverted by z and therefore this subgroup is abelian. As $[O_p(H_v), v] \neq 1$, there exists a v-minimal subgroup P_v in P_0 and $P_v = [P_v, v]$, so Lemma 2.1 (4) implies that $P_v = [P_v, v] = [P_v, z] \leq C_G(w)$. With the Pushing Down Lemma (3) it follows that $P_v \leq F(H_w)$ and therefore $p \in \pi_w$. Then Lemma 6.9, applied to P_v, gives that $N_G(P_v) \leq H_v$ and hence $H_w \hookrightarrow H_v$. We know that $2, p \in \pi_w$ and that p is odd, therefore $|\pi_w| \geq 2$ and Lemma 6.2 (3) forces $F_{\pi_w}(H_v) \leq H_w$. Therefore w centralises $F_{\pi_w}(H_v)$ and w inverts $F_{\pi'_w}(H_v)$ by the Infection Theorem (1). In particular w centralises P_0 and z inverts it, therefore v inverts P_0. The action of $\langle z, d \rangle$ on P_0 and Lemma 2.1 (4) give that
$$P_0 = \langle C_{P_0}(d), C_{P_0}(dz) \rangle$$
because z inverts P_0. If $P_1 := C_{P_0}(d) \neq 1$, then
$$P_1 = [P_1, z] \leq [H_d, z] \leq F(H_d)$$
because v and d are C-conjugate. As P_1 is contained in P_0 and P_0 is abelian, it follows that P_1 is centralised by U_v. So we deduce that $N_G(P_1) \leq H_v$ by Lemma 6.10 (and because $|\pi_v| \geq 2$). Thus $H_d \hookrightarrow H_v$ and the Infection Theorem (5), together with the fact that $E(H_v) = 1$ by (1), forces $H_v = H_d$. In particular $x \in H_v$. But $C_G(v)$ and $C_G(d)$ are both contained in H_v whence v and d are isolated in H_v by Lemma 13.5 (1). This is a contradiction.

Consequently $P_1 = 1$, hence $P_0 \leq C_G(dz) = H_w^x$ and therefore our v-minimal subgroup P_v is centralised not only by w, but also by dz and hence by $w \cdot dz = dv$. We let $H_{dv} := H_d^x$. As $P_v = [P_v, z] \leq [H_{dv}, z] \leq F(H_{dv})$ and $N_G(P_v) \leq H_v$, we have that $H_{dv} \hookrightarrow H_v$. Now we have to distinguish the cases where $v = a$ or $v = az$, and this is similar to our arguments in (2).

If $v = a$, then $dv = d^x$ is conjugate to v and Lemma 13.10 yields a contradiction.

If $v = az$, then dv is conjugate to $w = a$, the Infection Theorem (1) gives that $F_{\pi'_w}(H_v)$ is inverted by dv and Lemma 6.2 (3) yields that dv centralises $F_{\pi_w}(H_v)$. As U_v is centralised by w, we also have that $U_v = [U_v, z] \leq F(H_w)$ and hence $H_w \looparrowright H_v$, similarly $C_G(dz)$ infects H_v. But then it follows that $F_{\pi'_w}(H_v)$ is inverted by dv, by w and by dz, hence it is trivial, and $F^*(H_v) = F_{\pi_w}(H_v)$ is then centralised by w, forcing $w \in O_2(H_v)$. This is impossible by Lemmas 13.5 (1) and 13.1 (5). $\square$

(6) $F_{\pi'}(H_v) = 1$.

PROOF. Assume otherwise and let $F := F_{\pi'}(H_v)$. Recall that $U_v \leq F(C)$ and hence $U_v \leq F_\pi(H_v)$, so that $[F, U_v] = 1$. By (3) and the Infection Theorem (1) we have that z inverts F. The coprime action of $\langle d, z \rangle$ on F and Lemma 2.1 (4) yield that

$$F = \langle [C_F(d), z], [C_F(dz), z] \rangle.$$

If $F_1 := [C_F(d), z] \neq 1$, then F_1 is a non-trivial $U_v \langle v \rangle$-invariant subgroup of $F(H_v)$, but also a $U_d \langle d \rangle$-invariant subgroup of $F(H_d)$ (because $U_d \leq F_\pi(H_d)$ and $F_1 = [F_1, z] \leq F_{\pi'}(H_d)$). Then Lemma 6.10 forces $H_v = H_d$ and by Lemma 13.10 this is impossible.

It follows that $F = [C_F(dz), z]$, so with $H_{dz} = C_G(dz) = H_w^x$ we deduce that $F = [F, z] \leq F(H_{dz})$. Then $H_{dz} \looparrowright H_v$ because $1 \neq F \trianglelefteq H_v$ by assumption and because H_v is primitive by Corollary 4.8. As F is centralised by v, by (5), we also have that $F \leq C_G(dw)$. If we let $C_G(dw) \leq H_{dw} \max G$, then $F = [F, z] \leq F(H_{dw})$ and therefore $H_{dw} \looparrowright H_v$ as well. Now there are two cases again – if $v = a$, then w, dz and dw are conjugate and therefore $H_{dw} = C_G(dw)$. In particular $\pi_w = \pi(F(H_{dz})) = \pi(F(H_{dw}))$ and the Infection Theorem (1) and Lemma 6.2 (3) together yield that $dz \cdot dw$ centralises $F(H_v)$. This is false because $dz \cdot dw = v$ and $v \notin Z(H_v)$.

If $v = az$, then $w = a$ is conjugate to dz and to $wdz = dv$, so dw is conjugate to v and we can choose H_{dw} to be conjugate to H_v. Then Lemma 13.10 yields a contradiction. $\square$

We know from (3) that $C \looparrowright H_v$. Together with (1), (5) and the Infection Theorem (4) this forces $C = H_v$. This contradicts Lemma 13.1 (5). $\square$

THEOREM 13.23. *Suppose that Hypothesis* 10.1 *holds and let* $a \in C$ *be an involution that is chosen as in Lemma* 13.4. *Let* H_a *be a maximal subgroup of* G *containing* $C_G(a)$ *and let* H_{az} *be a maximal subgroup of* G *containing* $C_G(az)$. *Then one of the following holds:*

(1) $C_G(a) = H_a$ *and* $C_G(az) = H_{az}$.

(2) $C_G(a) = H_a$, $C_G(az) < H_{az}$ *with* H_{az} *having odd prime characteristic and one of the cases from List IV in the F*-Structure Theorem holds.*

(3) $C_G(a) < H_a$ *and* H_a *is of odd prime characteristic*, $C_G(az) = H_{az}$ *and one of the cases from List IV in the F*-Structure Theorem holds.*

(4) $C_G(a) < H_a$ *and* $C_G(az) < H_{az}$ *with* H_a *and* H_{az} *having the same odd prime characteristic.*

PROOF. Statement (1) is one of the possibilities, but we suppose now that it does *not* hold. The choice of a by hypothesis and some notation yield that Hypothesis 13.6 is satisfied. First suppose that $C_G(a) < H_a$. If $C_G(az) < H_{az}$, then we are in the situation of Hypothesis 13.15. Thus Theorem 13.17 and Lemma 13.18 are applicable and yield (4). Otherwise we have that $C_G(az) = H_{az}$ and we distinguish the two cases $C < M$ and $C = M$. If $C < M$, then we refer to Lemma 13.20 which gives (3). If $C = M$, then Lemmas 13.21 and 13.22 also give (3).

Similarly if $C_G(a) = H_a$ and $C_G(az) < H_{az}$, then we can again refer to Lemmas 13.20, 13.21 and 13.22 to see that (2) holds. □

CHAPTER 14

The Endgame

Our starting point in this chapter is a hypothesis building on Theorem 13.23 where we choose a suitable involution a centralising z and we set up notation that will be used throughout. In a series of results we first exclude the case where $C_G(a)$ is a maximal subgroup, and then we analyse the case with odd prime characteristic in order to reach a final contradiction.

HYPOTHESIS 14.1.
We suppose that Hypothesis 13.6 holds and for simplification, we set $b := az$.

We recall that, by Lemma 13.7, Hypothesis 14.1 implies Hypothesis 6.6.

LEMMA 14.2. *Suppose that Hypothesis 14.1 holds and that $r_2(G) = 2$. Then there exists an odd prime r such that, for all involutions $t \in C$ with $t \neq z$, the centraliser $C_G(t)$ lies in a maximal subgroup H_t of G of characteristic r.*

PROOF. Our hypothesis implies that Hypothesis 13.6 holds and that $F^*(\overline{C})$ is as in the first case on List II in the F*-Structure Theorem. Then it follows that all involutions in C distinct from z are conjugate (as can be seen in C) and therefore we may suppose that t is our involution a. By Theorem 13.23 we need to exclude the case that $C_G(a) = H_a$. We assume therefore, by way of contradiction, that $C_G(a) = H_a$. We know that $z \notin Z(H_a)$ by Lemma 13.1 (5) and then Lemma 4.2 (6) yields an odd prime $p \in \pi_a$ such that $[O_p(H_a), z] \neq 1$. Let $P \in \mathrm{Syl}_p(H_a, V)$ (with Lemma 13.5 (8)) and let $T \in \mathrm{Syl}_2(C_G(P))$. As $C_G(P)$ is z-invariant, we may suppose that T is z-invariant by Lemma 3.11. Then z centralises T by Lemma 3.1 (2). Moreover $a \in C_G(P) \leq H_a$, so we see that a is central in $C_G(P)$ and hence $a \in Z(T)$. The conjugacy of a and b yields that Case (1) from Lemma 6.11 holds, so there exists an involution distinct from a in $Z(T)$. Then $2 \leq r(Z(T)) \leq r_2(G) = 2$. But then it follows that $z \in Z(T) \leq T \leq C_G(P) \leq C_G(O_p(H_a))$, which is a contradiction. □

LEMMA 14.3. *Suppose that Hypothesis 14.1 holds and that $C_G(b)$ is a maximal subgroup. Then $C = M$.*

PROOF. Assume that $C < M$ and let $U_0 := [C_U(b), z]$.
Case 1: *a and b are conjugate.*

Then Lemma 13.9 implies that M has characteristic p. We also know that $C_G(a)$ and $C_G(b)$ are maximal subgroups and we let $x \in C$ be such that $b^x = a$. If $U_0 = 1$, then
$$1 = U_0^x = ([C_U(b), z])^x = [C_U(a), z]$$

and this means that $C_U(a)$ and $C_U(b)$ are both contained in $C_U(z)$. With Lemma 2.1 (4) this forces $U = C_U(z)$, which is a contradiction. Hence $U_0 \neq 1$. Lemma 2.1 (2) implies that $[U_0, z] \neq 1$ and hence $[U_0, a] \neq 1$. In particular we know that $[U, a] \neq 1$. We let H be a maximal subgroup of G containing $N_G(U_0)$, with Lemma 13.1 (1), and we assume that $H \neq M$.

(1.1) M and H_b infect H and b inverts $F_{\pi_b'}(H)$ and centralises $F_{\pi_b}(H)$.

PROOF. We have that $M \looparrowright H$ because $U_0 \leq F(M)$. With the Pushing Down Lemma (2), we see that
$$U_0 \leq [O_p(M) \cap H_b, z] \leq O_p(H_b)$$
and hence $H_b \looparrowright H$ as well. The Infection Theorem (1) implies that $F_{\pi_b'}(H) \cap H_b = 1$ whence $F_{\pi_b'}(H)$ is inverted by b. Also, the set π_b contains at least two distinct primes, namely 2 and p, and therefore Lemma 6.2 (3) yields that $F_{\pi_b}(H) \leq H_b = C_G(b)$. □

(1.2) $E(H) \leq H_b$.

PROOF. Assume that $E(H) \not\leq H_b$. As $C_G(V) \leq N_G(U_0) \leq H$, Lemma 13.14 (3) yields that H has a component L such that $L/O(L)$ is isomorphic to the simple component of $\overline{C}$ and such that $a \in L$. But then $[F(H), a] \leq [F(H), L] = 1$ which means, by (1.1), that a and b centralise $O_p(H)$. Hence $O_p(H) \leq C$. Then the Pushing Down Lemma (2) implies that
$$U_0 \leq [O_p(M) \cap H, z] \leq O_p(H) \leq C,$$
which is a contradiction. □

Now b centralises $E(H)$ and $F_{\pi_b}(H)$ by (1.2) and (1.1) and it inverts $F_{\pi_b'}(H)$ by (1.1). So we see that $[H, b]$ centralises $F^*(H)$ and hence $[H, b] \leq Z(F(H))$. Moreover we recall that $M \looparrowright H$ and that M is of characteristic p. In particular, with the Infection Theorem (1), it follows that $F_{p'}(H)$ is inverted by z. As $p \in \pi_b$, this implies that $F_{\pi_b'}(H)$ is inverted by z and by b and hence centralised by a. Now we let $P \in \mathrm{Syl}_p(H_b, V)$ (with Lemma 13.5 (8)).

(1.3) There exists a conjugate v of b commuting with b and such that, with $H_v := C_G(v)$, we have the following:
- $H_v \looparrowright H$ and
- $H = C_G(vb)$.

PROOF. First we note that $1 \neq U_0 \leq [O_p(H_b), z] \leq P$. Since a and b are conjugate, Lemma 6.11 (1) yields that $N_G(P) \not\leq H_b$. Let $T \in \mathrm{Syl}_2(C_G(P))$. We may suppose that T is z-invariant by Lemma 3.11, because $C_G(P)$ is z-invariant. With Lemma 6.11 (1), let $c \in N_G(P) \cap N_G(T) \cap C$ and $v \in Z(T)$ be such that $b^c = v \neq b$. Let $H_v := H_b^c (= C_G(v))$. Then $U_0 \leq P \leq H_v$ and, again with the Pushing Down Lemma (2), we deduce that
$$U_0 \leq [U \cap H_v, z] \leq O_p(H_v).$$
Therefore $H_v \looparrowright H$ and as in (1.1) it follows that $F_{\pi_b'}(H)$ is inverted by v and that $F_{\pi_b}(H)$ is contained in H_v and hence centralised by v.

We recall that $E(H) \leq H_b$, so that $E(H)$ is centralised by b and by z, hence by a as well. If we check the possibilities from Lists I, II and III for quasi-simple subgroups of $F^*(\overline{C})$ that are centralised by $\overline{V}$, then we only see groups that are isomorphic to $SL_2(q)$ with a suitable power q of some odd prime and such that their central involution is $\overline{z}, \overline{a}$ or $\overline{b}$. As $z \notin Z(H)$, we know from Lemma 4.2 (6) that $z \notin E(H)$. Therefore, if $E(H) \neq 1$, then H possesses a unique component and this component has a central involution that is conjugate to a (and then to b). We recall that $H_b \looparrowright H$ and then Lemma 13.10 forces $H = H_b$. Thus $M \looparrowright H$ by (1.1) and, as $\operatorname{char}(M) = p$, the Infection Theorem (1) gives that $F_{p'}(H) \cap M = 1$. This is false because $b \in O_2(H) \cap M \leq F_{p'}(H) \cap M$. We conclude that $E(H) = 1$ and therefore $F^*(H) = F(H)$ is centralised by bv. With H_{bv} denoting a maximal subgroup of G containing $C_G(bv)$, this means that $F^*(H) \leq H_{bv}$ and in particular $H \looparrowright H_{bv}$. Also, we see that $U_0 \leq F(H) \leq H_{bv}$ whence the Pushing Down Lemma (2) implies that $U_0 \leq O_p(H_{bv})$. So conversely we have that $H_{bv} \looparrowright H$.

With the Infection Theorem (3), our claim follows or H and H_{bv} are distinct and have both characteristic p. But $H_b \looparrowright H$ by (1.1) and then the Infection Theorem (2) forces H_b to have characteristic p as well. This is impossible because $O_2(H_b) \neq 1$. Consequently $C_G(bv) \leq H$, then Lemma 13.5 (1) implies that bv is isolated in H and we saw above that $bv \in C_H(F^*(H)) = Z(F(H))$. This forces $bv \in Z(H)$. □

With the notation from (1.3) we have that bv is central in H and that H_b, H_v and M infect H. We recall that b and v are centralised by z, so $z \in H$ and therefore $O_2(H)$ is centralised by z. In particular $O_2(H) \cap M \neq 1$. However, $M \looparrowright H$ whence the Infection Theorem (1) implies that $F_{\pi'}(H) \cap M = 1$. This is a contradiction because $2 \notin \pi$ and hence $O_2(H) \leq F_{\pi'}(H) \cap M$.

This last contradiction comes from the assumption that $H \neq M$ (before (1.1)), so we have established that $N_G(U_0) \leq M$ and in particular $H_b \looparrowright M$. But M is of characteristic p and hence H_b is of characteristic p by the Infection Theorem (2). This is a contradiction.

Case 2: a and b are not conjugate.

As before let $U_0 := [C_U(b), z]$. If $E(M) \neq 1$, then Lemma 13.9 yields that there exists an odd number $q \geq 11$ such that $E(M) \simeq PSL_2(q)$. Therefore Hypothesis 7.1, and more specifically the hypothesis of Lemma 7.6, is satisfied. The lemma gives that

$$U \leq \bigcap_{g \in G} M^g,$$

but this contradicts Lemma 13.1 (1).

We conclude that $E(M) = 1$ and then our assumption that $C < M$ and $C_G(b) = H_b$, together with Lemma 13.19, yields that $[U, a] \neq 1$. If $U_0 = 1$, then $C_U(b) \leq C_U(z)$ and it follows from Lemma 2.1 (4) that

$$U = [U, z] \leq [U, b] \cap C_G(a),$$

contrary to the earlier remark that $[U, a] \neq 1$. Therefore $U_0 \neq 1$. As before we let $N_G(U_0) \leq H \max G$, so that $M \looparrowright H$. With the Pushing Down Lemma (2), it follows that $U_0 \leq O_p(H_b)$ and hence $H_b \looparrowright H$. We also see, with the same result, that $U_0 \leq [F(M) \cap H, z] \leq F(H)$ which yields that $U_0 \leq O_p(H)$. With Lemma 13.5 (8) let $P \in \mathrm{Syl}_p(H_b, V)$ and let $b \in T \in \mathrm{Syl}_2(C_G(P))$. We may suppose that T is z-invariant by Lemma 3.11, applied to $C_G(P)\langle z \rangle$. If $N_G(P)$ is not contained in H_b, then with Lemma 6.11 (1) there exists an element $x \in N_G(P) \cap N_G(T) \cap C$ such that $d := b^x \in Z(T)$ and $d \neq b$. This leads to the next step in the proof, so we keep this notation.

(2.1) If $N_G(P)$ is not contained in H_b, then $H = C_G(bd)$.

PROOF. We note that T is z-invariant and hence centralised by z (by Lemma 3.1 (2)), but that z is not contained in T because $1 \neq U_0 \leq [P, z]$. Together with Theorem **D** this implies that $\Omega_1(Z(T)) = \langle b, d \rangle$. Let $H_d := H_b^x$. Then $U_0 \leq P = P^x \leq H_d$ and, with the Pushing Down Lemma (2), it follows that

$$U_0 = [U_0, z] \leq [O(F(M)) \cap H_d, z] \leq F(H_d).$$

Therefore $U_0 \leq O_p(H_d)$ and $H_d \looparrowright H$ as well. The Infection Theorem (1) and Lemma 6.2 (3), together with the fact that b and d are conjugate, yield that b and d invert $F_{\pi_b'}(H)$ and centralise $F_{\pi_b}(H)$. Thus bd centralises $F(H)$ which implies that $U_0 \leq F(H) \leq C_G(bd)$. We know that a lies in a 2-component E of C such that $\overline{E}$ is isomorphic to $PSL_2(q)$ with some odd prime power $q \geq 11$, by our hypothesis in this case. Then also $dz = b^x z = (bz)^x = a^x \in E$ because $x \in C$. Therefore $\langle a, dz \rangle$ is a fours group and bd is conjugate to a and to dz. Let $C_G(bd) \leq H_{bd} \max G$. Then we may choose H_{bd} to be conjugate to H_a. Another application of the Pushing Down Lemma (2) yields that

$$U_0 = [U_0, z] \leq [O(F(M)) \cap H_{bd}, z] \leq O_p(H_{bd})$$

and hence $H_{bd} \looparrowright H$.

Assume that $E(H) \neq 1$. We know that $C_G(V) \leq H$ and hence Lemma 13.14 yields that either $E(H)$ is centralised by b or H possesses a component L such that $L/O(L)$ is simple and $a \in L$. In the latter case we see, since $H_b \looparrowright H$ and $p \in \pi_b$, that $O_p(H)$ is centralised by b and by a, and therefore $U_0 \leq C$. But this is impossible. Therefore $E(H) \leq H_b$. As a and b are not conjugate in this case, Lemma 13.5 (1) gives that one of the cases from List IV holds and therefore the only way that H can have a component is that this component has z as a central involution. This is impossible because $H \neq C$. We conclude that $E(H) = 1$.

Then $F^*(H) = F(H)$ is centralised by bd, it follows that $F^*(H) \leq H_{bd}$ and hence $H \looparrowright H_{bd}$. If H has characteristic p, then we recall that $U_0 \leq F(H_b)$ and hence $H_b \looparrowright H$. So the Infection Theorem (2) yields that H_b has characteristic p as well, which is impossible. As H and H_{bd} infect each other, the Infection Theorem (3) gives that $H =$

H_{bd}. In particular H_{bd} does not have characteristic p and Theorem 13.23 implies that $C_G(bd) = H_{bd}$. □

(2.2) $N_G(P) \leq H_b$.

PROOF. Otherwise we know from (2.1) that $N_G(U_0) \leq H_{bd} = C_G(bd)$. The choice of a and b implies that bd is conjugate to a and hence that $C_G(bd)$ is conjugate to H_a. We recall that $U_0 \leq O_p(M)$ and hence $M \hookrightarrow H_{bd}$, and we also recall that $H_{bd} = C_G(bd)$ and hence $2 \in \pi(F(H_{bd}))$. Then the Infection Theorem (1) forces $F_{\pi'}(H_{bd}) \cap M = 1$. This is a contradiction because π consists of odd primes and hence $bd \in O_2(H_{bd}) \leq F_{\pi'}(H_{bd}) \cap C$. □

Now we have that $N_G(P)$ is contained in H_b and Lemma 6.11 (2) yields that every z-invariant p-subgroup of $C_G(a)$ is centralised by z. This forces $[C_U(a), z] = 1$ and therefore $C_U(a) \leq C_U(z)$. But a is contained in a fours group A_0 of C such that all involutions of A_0 are C-conjugate, so Lemma 2.1 (4) implies that

$$U = \langle C_U(a_0) \mid a_0 \in A_0 \rangle \leq C_U(z)$$

whence $U \leq C$, which is a contradiction.

□

THEOREM 14.4. *Suppose that Hypothesis* 14.1 *holds and that* $C_G(b)$ *is a maximal subgroup. Then* $O(F(C)) \cap H_b = 1$.

PROOF. Assume otherwise. Lemma 14.3 yields that $C = M$ and it follows with Lemma 6.12 that $[H_b, z] \leq F_{\pi'}(H_b)$. Our hypotheses that $O(F(C)) \cap H_b \neq 1$ and that $C_G(b) = H_b$ imply that some element from $O(F(C))$ is centralised by b (and, of course, by z) and hence by a. Therefore $O(F(C)) \cap C_G(a) \neq 1$ and in particular $O(F(C)) \cap H_a \neq 1$ as well. Then $[H_a, z] \leq F_{\pi'}(H_a)$, again by Lemma 6.12. We use this fact frequently and therefore refer to the following (also with applications to conjugates of a or b):

$$[H_a, z] \leq F_{\pi'}(H_a) \text{ and } [H_b, z] \leq F_{\pi'}(H_b) \qquad (*).$$

Case 1: a and b are conjugate.

From Lemma 13.1 (5) we know that $H_a \neq C$ and then, by Lemma 4.2 (6), that there exists an odd prime $p \in \pi_a$ such that $P_0 := [O_p(H_a), z] \neq 1$. In particular $|\pi_a| \geq 2$. With Lemma 13.1 (1) we let $N_G(P_0) \leq H \max G$. Moreover let $P \in \mathrm{Syl}_p(H_a, V)$ by Lemma 13.5 (8) and let $a \in T \in \mathrm{Syl}_2(C_G(P))$. We may suppose that T is z-invariant by Lemma 3.11, applied to $C_G(P)\langle z \rangle$. As a and b are conjugate, Lemma 6.11 (1) holds and there exist an involution v and an element $c \in N_G(Q) \cap N_G(T) \cap C$ such that $v := a^c \neq a$. Let $H_v := H_a^c$.

(1.1) H_a and H_v infect H. The involutions a and v invert $F := F_{\pi'_a}(H)$ and centralise $F_{\pi_a}(H)$. Moreover $H \neq C$.

PROOF. As $O_p(H_a) \neq 1$ and H_a is primitive by Corollary 4.8, we know that

$$T \leq C_G(P) \leq C_G(O_p(H_a)) \leq H_a = C_G(a)$$

and hence $a \in Z(T)$. We also have that $P_0 \leq P \leq H_v$. It follows from conjugacy and $(*)$ that $[H_v, z] \leq F_{\pi'}(H_v)$ and therefore $P_0 \leq F(H_v)$. Now $H_a \hookrightarrow H$, also $H_v \hookrightarrow H$ and the Infection Theorem (1) yields that a and v invert F as stated. Lemma 6.2 (3) gives that a and v centralise $F_{\pi_a}(H)$.

If $H = C$, then in particular $P_0 \leq C$ which forces $P_0 = 1$ (by Lemma 2.1 (2)). But this is a contradiction. $\square$

(1.2) If $H \neq H_a$, then $E(H) = 1$.

PROOF. The subgroup P_0 is $C_G(V)$-invariant and so $C_G(V) \leq H$. Then Lemma 13.14 (3) implies that $E(H) \leq H_a$ or that one of the cases from List IV holds. As a and b are conjugate, Lemma 13.5 (2) yields that $F^*(\overline{C})$ is as on List I, II or III and therefore $E(H) \leq H_a$. Now we assume that $E(H) \neq 1$. By (1.1), no component of H can have z as central involution, so only the cases from List II remain and $E(H)$ has a unique component, with central involution a or b. If the central involution is b, then $H = H_b$ and hence $P_0 \leq H_b$. But then P_0 is centralised by a and by b, hence by z, and this is a contradiction. Thus the central involution in the component of H is a and this implies that $H = H_a$, contrary to our assumption. We conclude that $E(H) = 1$. $\square$

Assume that $H \neq H_a$. Then it follows from (1.2) that $E(H) = 1$ and thus, with (1.1), we see that av centralises $F(H) = F^*(H)$. The choice of a and the F*-Structure Theorem yield that a is C-conjugate to av and hence $H_{av} := C_G(av)$ is a maximal subgroup of G.

(1.3) If $H \neq H_a$, then $H = C_G(av)$.

PROOF. We saw above that $F^*(H) \leq H_{av}$ and hence we know that $av \in C_H(F^*(H)) = Z(F(H))$. This implies that $av \in O_2(H)$. Moreover $O_2(H)$ is centralised by a, v and z, but $z \notin O_2(H)$. If $a \in O_2(H)$, then a centralises $F(H)$ and hence $F(H)$ is a π_a-group. But $E(H) = 1$ and then the Infection Theorem (4) forces $H = H_a$, contrary to our assumption. The same argument yields that $v \notin O_2(H)$ because otherwise $H = H_v$ by infection and then, as H_a and H_v are conjugate, the Infection Theorem (5) implies that $H_a = H_v = H$, which contradicts our assumption once more. We know that G has at most rank 3 by Theorem **D** and so it follows that $O_2(H)$ has at most rank 2. As $\langle a, v, z \rangle$ centralises $O_2(H)$, we see that either av is the only involution in $O_2(H)$ or that $\langle av, vz \rangle \leq O_2(H)$. Thus if av is not central in H, then av, vz and b are contained in $O_2(H)$ and conjugate in H. In particular b centralises $F^*(H)$, just as av does, and therefore

$$P_0 = [P_0, z] = [P_0, b] \leq [H, b] \leq C_H(F^*(H)) = Z(F(H)).$$

But then $P_0 = [P_0, z] \leq [H_{av}, z] \leq F_{\pi'}(H_{av})$, by $(*)$, because a and av are conjugate. Thus $H_{av} \hookrightarrow H$ and it follows that $F(H)$ is a π_a-group. We recall that $\pi_a = \pi(F(H_{av}))$ and hence the Infection Theorem (4) gives that $H = H_{av}$. $\square$

If $H \neq H_a$, then $H = H_{av}$ by (1.3) and hence H_v and H_a infect H_{av}. These subgroups are conjugate and $E(H) = 1$ by (1.2), so the Infection Theorem (5) gives that $H_a = H_{av} = H_v$. But this is impossible by Lemma 13.3 (1). This contradiction comes from our assumption that $O(F(C)) \cap H_a \neq 1$, so the Lemma is proved in the case where a and b are conjugate.

Case 2: a and b are not conjugate.

It will be important here that, by Lemma 13.5 (2), one of the cases from List IV holds and therefore all involutions in C are conjugate to a or to b. Then a is contained in a 2-component of C and, in this 2-component, in a fours group with involutions a, v, av that are all conjugate in C. Let $C_G(v) \leq H_v \max G$ and $C_G(av) \leq H_{av} \max G$. These maximal subgroups are conjugate to H_a. Conjugacy implies that $[H_v, z] \leq F_{\pi'}(H_v)$ and $[H_{av}, z] \leq F_{\pi'}(H_{av})$, we keep referring to this by $(*)$ as before. Moreover set $w := vz$ and let $C_G(w) \leq H_w \max G$, with H_w chosen to be conjugate to H_b.

We quote Lemmas 13.1 (5) and 4.2 (6) once more and this time choose an odd prime $p \in \pi_b$ such that $P_0 := [O_p(H_b), z] \neq 1$. The objective of the following steps is to prove that

$$N_G(P_0) \leq H_b.$$

We let $N_G(P_0) \leq H \max G$ and $F := F_{\pi_b'}(H)$, and we assume that $H \neq H_b$ and $F \neq 1$. It is worth noticing here that $H \neq C$ because z does not centralise P_0.

(2.1) z inverts F.

PROOF. First we set up some notation. Let $r \in \pi(F)$, let $R := O_r(H)$ and $R_0 := C_R(z)$. By way of contradiction we assume that $R_0 \neq 1$.

(a) $C \looparrowright H$, $E(H) = 1$ and $[H, z] \leq Z(F(H))$.

PROOF. We have that $H_b \looparrowright H$ and $C_C(b) \leq H$. With the Infection Theorem (1) and Lemma 6.2 (3), we see that b inverts F and centralises $F_{\pi_b}(H)$. Also, we know that $2 \in \pi_b$ and hence F is a $C_C(b)$-invariant subgroup of odd order. Now we suppose that $R_0 \neq 1$. Then R_0 is a non-trivial nilpotent $C_C(b)$-invariant $2'$-subgroup of C and Lemma 13.12 yields that $1 \neq R_0 = [R_0, b] \leq F(C)$. Hence $r \in \pi$. This implies that $R_0 = R$ with Lemma 4.10. Moreover $R = R_0 \leq O_r(C)$ means that $C \looparrowright H$ as stated, because $N_G(R) = H$ by Corollary 4.8. We know that z centralises $E(H)$ by Lemma 4.2 (5), moreover z centralises $F_\pi(H)$ by Lemma 4.10 and z inverts $F_{\pi'}(H)$ by the Infection Theorem (1) because $C \looparrowright H$. Therefore $[H, z] \leq C_H(F^*(H)) = Z(F(H))$. In particular $P_0 \leq Z(F(H))$.

As P_0 is $C_G(V)$-invariant, Lemma 13.14 yields that $E(H) \leq H_b$ or that H has a unique component and that a is contained in this component. In the latter case we see that a centralises $F(H)$ and hence P_0 is centralised by a and by b, hence by z, and this is impossible. It follows that $E(H) \leq H_b$. But this is impossible in the cases from List IV (because $z \notin E(H)$), so we actually see that $E(H) = 1$. □

(b) Let $X := C_{O_p(H)}(w)$. Then z inverts X and $X \leq C_G(av)$. Moreover if $X \neq 1$, then $N_G(X) \leq H$.

PROOF. Suppose that $X \neq 1$ because otherwise the first statement in (b) is clear. With Lemma 13.1 (1) let $N_G(X) \leq H_1 \max G$. Then $H \looparrowright H_1$. First we note that $p \in \pi' \cap \pi_b$, applying Lemma 4.10, because z does not centralise P_0. Then (a) and the Infection Theorem (1) imply that z inverts $O_p(H)$ whence

$$O_p(H) = [O_p(H), z] \leq [H_b, z] \leq F(H_b)$$

by (*). Thus $O_p(H) \leq P_0$. Conversely $P_0 = [P_0, z] \leq O_p(H)$ by (a) and therefore $P_0 = O_p(H)$. This also yields that $O_p(H)$ is abelian and hence $F^*(H)$ centralises X. Now we know that $X \leq F(H_b)$ and therefore $H_b \looparrowright H_1$. As X is inverted by z and centralised by w and by b (and therefore by $bw = av$, too), we see that

$$X = [X, z] \leq [H_w, z] \leq F(H_w),$$

again by (*), and similarly $X \leq F(H_{av})$. Now it follows that $H_w \looparrowright H_1$ and $H_{av} \looparrowright H_1$. By the Infection Theorem (1), the involutions b, w and av ($= bw$) invert $F_{\pi'_b}(H_1)$, so this subgroup must be trivial. Moreover $F^*(H) \leq C_G(X) \leq H_1$, in particular $F = [F, b] \leq H_1$. Lemma 6.2 (3) gives that $F(H_1) = F_{\pi_b}(H_1)$ is centralised by b, hence by $[H_1, b]$ and this implies that $F = [F, b]$ commutes with $F(H_1)$. We deduce that $F(H_1) \leq N_G(F) = H$ (by Corollary 4.8). Next we recall that $[H, z] \leq Z(F(H))$ by (a), in particular $P_0 \leq F(H) \leq H_1$. Therefore $E(H_1)$, which is centralised by z by Lemma 4.2 (5), is centralised by $P_0 = [P_0, z]$. Thus $E(H_1) \leq C_G(P_0) \leq H$, and since we already established that $F(H_1) \leq H$, we conclude that $F^*(H_1) \leq H$. Consequently $H_1 \looparrowright H$ and the Infection Theorem (3) yields that $H_1 = H$ as stated, because H is not of prime characteristic. □

(c) If $C_{O_p(H)}(w) \neq 1$, then $H = H_{av}$.

PROOF. Let $X := C_{O_p(H)}(w)$ and suppose that $X \neq 1$. Then $X \leq C_G(av)$ and $N_G(X) \leq H$ by (b). It follows with (*) and (b) that $X = [X, z] \leq F(H_{av})$ and therefore $H_{av} \looparrowright H$. As $X \leq H_w$, we also have with (*) that $X = [X, z] \leq F(H_w)$ and hence $H_w \looparrowright H$. Moreover $H_b \looparrowright H$ (from the start) and then the Infection Theorem (1) and Lemma 6.2 (3) imply that b and w centralise $F_{\pi_b}(H)$ and invert F. Therefore $bw = av$ centralises $F(H) = F^*(H)$ by (a). In particular $F^*(H) \leq H_{av}$ whence $H \looparrowright H_{av}$. Now the Infection Theorem (3), together with the fact that H is not of characteristic p, yields that $H = H_{av}$. □

(d) w inverts $O_p(H)$.

PROOF. Assume that $X := C_{O_p(H)}(w) \neq 1$. Then $H = H_{av}$ by (c) and, as av is conjugate to a, Theorem 13.23 yields that there are two cases to consider:
av is central in H_{av} or H_{av} has odd prime characteristic.
In the first case we see that

$$F(H_b) \leq N_G(P_0) \leq H = C_G(av),$$

so $F(H_b)$ is centralised by b, by $av = bw$ and hence by w as well. With Lemma 13.8 this gives that $F^*(H_b) \leq H_w$ and thus $H_b \looparrowright H_w$. With the same lemma, conjugacy and the Infection Theorem (5), we conclude that $H_b = H_w$, which contradicts Lemma 13.10. In the second case we must have that $\operatorname{char}(H_{av}) = p$ because $1 \neq X \leq O_p(H_{av})$. But still $H_b \looparrowright H = H_{av}$, so the Infection Theorem (2) forces H_b to be of characteristic p as well, and this is impossible because $2 \in \pi_b$. This contradiction shows that $X = 1$. □

(e) $O_p(H) = [O_p(H), z] \leq C_G(v)$.

PROOF. We know by (d) that $w = vz$ inverts $O_p(H)$. We have already noticed (with Lemma 4.10) that $p \in \pi'$ because z does not centralise P_0. As $C \looparrowright H$ by (a), the Infection Theorem (1) forces $O_p(H)$ to be inverted by z as well and hence to be centralised by v. This yields the statement. □

(f) If $P \in \operatorname{Syl}_p(H_b, V)$, then $N_G(P) \not\leq H_b$.

PROOF. Assume that $N_G(P) \leq H_b$. Then Lemma 6.11 (2) applies and in particular every z-invariant p-subgroup of $C_G(a)$ is centralised by z. By conjugacy, every z-invariant p-subgroup of $C_G(v)$ is centralised by z, so (e) forces

$$O_p(H) = [O_p(H), z] = [O_p(H), z, z] = 1.$$

This is impossible because $1 \neq P_0 \leq O_p(H)$. □

(g) There exists a C-conjugate d of b such that $d \neq b$ and such that d centralises $F_{\pi_b}(H)$ and inverts F.

PROOF. Let $P \in \operatorname{Syl}_p(H_b, V)$, by Lemma 13.5 (8) and let $T \in \operatorname{Syl}_2(C_G(P))$. As $C_G(P)$ is z-invariant, we may suppose that T is z-invariant by Lemma 3.11. Then (f) and Lemma 6.11 (1) imply that $Z(T)$ contains a fours group with b in it, but not z, and with some involution d in this fours group being distinct from b and conjugate to it with an element from $N_G(T) \cap N_G(P) \cap C$. Of course $C_G(P) \leq C_G(P_0) \leq H$, so if we let $H_d := C_G(d)$, then this maximal subgroup is conjugate to H_b and contains P, hence P_0. This means that $P_0 = [P_0, z] \leq F(H_d)$ by (∗) and therefore $H_d \looparrowright H$ as well.

Then Lemma 6.2 gives that d centralises $F_{\pi_b}(H)$ and the Infection Theorem (1) yields that d inverts F. □

(h) $H = H_v$.

PROOF. Let d be an involution as in (g). Then bd centralises $F(H) = F^*(H)$ whence $F^*(H) \leq C_G(bd)$. Here we should note that, as b and d are C-conjugate, it follows that bd is conjugate to a. With H_{bd} being a maximal subgroup containing $C_G(bd)$, we may choose H_{bd} to be conjugate to H_a and then $P_0 \leq F(H_{bd})$, again by (∗). Hence $H \looparrowright H_{bd} \looparrowright H$. As $\pi(F(H))$ contains the distinct primes r and p, the Infection Theorem (3) forces $H = H_{bd}$. In particular H_{bd} is not of prime characteristic (because $H_b \looparrowright H_{bd}$ now) and therefore H_a is not of prime characteristic. We deduce that $C_G(a) = H_a$ by Theorem 13.23 and then $C_G(v) = H_v$, by conjugacy. But (e) implies that $P_0 \leq F(H_v)$ (by (∗)) whence $H_v \looparrowright H = H_{bd}$. Finally Lemma 13.10 tells us that $H_v = H_{bd} = H$. □

We return to the prime $r \in \pi(F) \cap \pi'$ from the beginning of the proof of (2.1). As H_b and C infect H (by choice of H and by (a)), the Infection Theorem (1) yields that $O_r(H)$ is inverted by z and by b, hence centralised by a. Thus (∗) implies that

$$O_r(H) = [O_r(H), z] \leq F(H_a).$$

It follows that $H_a \looparrowright H = H_v$ because $N_G(O_r(H)) = H$ by Corollary 4.8. As a and v are conjugate and $E(H) = 1$ by (a), the Infection Theorem (5) forces $H_v = H_a$. But this is impossible by Lemma 13.3 (2).

This means that z inverts F. □

(2.2) $H_a \looparrowright H$ and $F_{\pi'_a}(H)$ is inverted by z.

PROOF. As H_b infects H, the Infection Theorem (1) and (2.1) above yield that b and z invert F and hence a centralises F. Our assumption that $F \neq 1$ and the fact that $F = [F, z] \leq F(H_a)$ with (∗) yields that $H_a \looparrowright H$, because H is primitive. In particular $F_{\pi'_a}(H)$ is inverted by a, with the Infection Theorem (1).

If H_a has odd prime characteristic, then by our main hypothesis, we have an a-minimal subgroup $U_a \leq O_{r_a}(H_a)$. As $F \leq C_{F(H_a)}(a) \leq C_G(U_a)$ by Lemma 6.7, we see that $U_a \leq C_G(F) \leq H$. Hence every normal subgroup of H which is inverted by a is centralised by U_a and therefore lies in H_a with Lemma 6.9. Together with the Infection Theorem (1), this forces $F(H)$ to be an r_a-group and this is impossible. Therefore $C_G(a) = H_a$ by Theorem 13.23. Now we have symmetry between a and b and therefore the previous arguments are applicable. We assume that $r \in \pi'_a$ is such that $R_0 := C_{O_r(H)}(z) \neq 1$, as we have done in (2.1(a)). Then from Lemma 13.12 we derive that $R_0 \leq O_r(C)$. In particular $r \in \pi$ and $R_0 = O_r(H)$. We see that $C \looparrowright H$ and if H has a component, then it contains a and thus $[F(H), a] = 1$. But $1 \neq O_r(H) = [O_r(H), b] = [O_r(H), a]$, which is a contradiction. These statements correspond to (a). Then we continue as for (b) and set $X := C_{O_p(H)}(w)$. If $X \neq 1$, then $N_G(X) \leq H$. This proof only needed $E(H) = 1$. We argue further that $F \neq 1$ and $R_0 \neq 1$ force $X = 1$, so X is inverted by z and by w, hence centralised by v which is conjugate to a. This corresponds to (e). Thus it follows as in (2.1) that z inverts $F_{\pi'_a}(H)$. □

(2.3) $[H, z] \leq Z(F_{\pi'}(H))$ and $E(H) = 1$.

PROOF. With (2.1) and (2.2) we see that, for all primes q in $\pi(F(H))$, we have that $q \in \pi'_b$ (so z inverts $O_q(H)$) or that $q \in \pi'_a$ (and again z inverts $O_q(H)$) or that q is in $\pi_a \cap \pi_b$ in which case $O_q(H)$ is centralised by a and b and hence by z. Now z centralises $E(H)$ and centralises or inverts $O_q(H)$ for every $q \in \pi(F(H))$. We deduce that $[H, z] \leq Z(F_{\pi'}(H))$.

For the second statement we quote Lemma 13.14. As $C_G(V) \leq H$, it yields that $E(H) \leq H_b$ or that H has a unique component and that this component contains a. In the latter case it follows that a centralises $F(H)$ and, as $P_0 \leq F(H)$ by the first paragraph, it follows that $[P_0, a] = 1$. But then z centralises P_0, which is a contradiction. Thus $E(H) \leq H_b$ and, looking at the cases from List IV, this is only possible if z is contained in (and hence central in) $E(H)$. But this is impossible because $H \neq C$. It follows that $E(H) = 1$. □

We know that z does not centralise $F(H_a)$ either, so we find a prime $p_a \in \pi_a$ such that $Q_0 := [O_{p_a}(H_a), z] \neq 1$ and we can play the same game with Q_0 instead of P_0.

(2.4) $C_G(a) = H_a$ and $N_G(Q_0) \leq H_a$.

PROOF. If $C_G(a) \neq H_a$, then we have that $F^*(H_a) = O_{p_a}(H)$, with Theorem 13.23 and our choice of p_a. We know that a centralises F and hence that $F = [F, z] \leq O_{p_a}(H_a)$ by $(*)$. Moreover F centralises U_a by Lemma 6.7. Then Lemma 6.9 implies that $U_a \leq H$. Also, the subgroup $F_{p'_a}(H)$ is inverted by a because $H_a \looparrowright H$. Therefore U_a centralises $F_{p'_a}(H)$, forcing $F(H)$ to be a p_a-group (first with Lemma 6.9 and then with the Infection Theorem (1)). Our hypothesis that $F \neq 1$ yields that $E(H) = 1$, by (2.3), so it follows that H has characteristic p_a. But this is impossible by the Infection Theorem (2) because $H_b \looparrowright H$. Thus we deduce that $C_G(a) = H_a$.

Now let $N_G(Q_0) \leq H_1 \max G$ and assume that $H_1 \neq H_a$. If we assume that $F_1 := F_{\pi'_a}(H_1) \neq 1$, then we note that Lemma 13.14 (3) implies that $E(H_1) \leq H_a$. (Otherwise $F_1 = 1$ because a centralises it if a is in a component of H_1.) As $H_1 \neq C$, this forces $E(H_1) = 1$ (looking at the possibilities from List IV). We recall that $C_G(a)$ and $C_G(b)$ are both maximal subgroups now and therefore the arguments from (2.1) yield that F_1 is inverted by z and hence centralised by b. Thus $F_1 \leq F(H_b)$ by $(*)$ and we obtain, as in (2.2), that $H_b \looparrowright H_1$. That leads to the fact that $[H_1, z] \leq Z(F_{\pi'}(H_1))$ as in (2.3).

We combine this information – we know that $1 \neq F \leq F(H_a)$ and therefore $F \leq N_G(Q_0) \leq H_1$. This gives that $F = [F, z] \leq F(H_1)$

and thus $H_1 \looparrowright H$. The same argument, the other way around, gives that F_1 (which is inverted by a and by z, thus centralised by b) lies in $F(H_b)$ and therefore normalises P_0, giving

$$F_1 = [F_1, z] \leq [H, z] \leq F(H).$$

So $H \looparrowright H_1$ as well. Both these subgroups are not of prime characteristic, so the Infection Theorem (3) forces $H = H_1$. This implies that $F(H_a)$ and $F(H_b)$ are contained in a common maximal subgroup of G, namely H. But the fact that $[H_a, z] \leq F(H_a)$ yields that $\langle C_K(a) \rangle \leq F(H_a)$ and similarly $\langle C_K(b) \rangle \leq F(H_b)$. Therefore $C_K(a), C_K(b) \subseteq H$ and Lemma 13.1 (4) gives that $G \leq H$, which is a contradiction. □

(2.5) $Q_0 \not\leq H$ and $F = O_{p_a}(H)$.

PROOF. If $Q_0 \leq H$, then $Q_0 = [Q_0, z] \leq F(H)$ by (2.3). Thus we have that $H \looparrowright H_a$ by (2.4). But also $H_a \looparrowright H$ by (2.2) and $|\pi_a| \geq 2$, so it follows with the Infection Theorem (3) that $H = H_a$. Then P_0 is centralised by a and by b, hence by z. This is impossible. Next we assume that $F \neq O_{p_a}(H)$. As

$$[O_{p_a'}(F), Q_0] \leq O_{p_a'}(F) \cap O_{p_a}(H_a) = 1,$$

we then see that $Q_0 \leq C_G(O_{p_a'}(F)) \leq H$ by Corollary 4.8 and this contradicts the first statement. Thus $F = O_{p_a}(H)$. □

(2.6) w centralises F.

PROOF. Let $P \in \mathrm{Syl}_p(H_b, V)$ and assume that $N_G(P) \not\leq H_b$. Then, by Lemma 6.11 (1), there exists a C-conjugate e of b centralising P, b and z. Let $H_e := C_G(e)$. We have that $P_0 \leq [H_e, z] \leq F(H_e)$ by $(*)$ whence $H_e \looparrowright H$. Applying the Infection Theorem (1) and Lemma 6.2 (3), it follows that F is inverted by b and by e and that $F_{\pi_b}(H)$ is centralised by b and e. With (2.3) we deduce that $F^*(H) = F(H)$ is centralised by be, an involution that is C-conjugate to a. Then (2.4) implies that $H_{be} := C_G(be)$ is a maximal subgroup of G and as $F^*(H) \leq H_{be}$, we now have that $H \looparrowright H_{be}$. But also $P_0 \leq [H_{be}, z] \leq F(H_{be})$ by $(*)$ and therefore $H_{be} \looparrowright H$. Our assumption that $F \neq 1$ guarantees that $F(H)$ is not a p-group and therefore the Infection Theorem (3) yields that $H = H_{be}$. But be is conjugate to a and $H_a \looparrowright H$ by (2.2), so Lemma 13.10 forces $H = H_a$. Then P_0 is centralised by b and by a, hence by z, and this is a contradiction. We conclude that $N_G(P) \leq H_b$ and that, with Lemma 6.11 (2), every z-invariant p-subgroup of $C_G(a)$ is centralised by z. The same statement holds for all involution centralisers $C_G(t)$ where t is C-conjugate to a. Now we look at the action of $\langle v, z \rangle$ on F. Lemma 2.1 (4) yields that

$$F = [F, z] = \langle [C_F(z), z], [C_F(v), z], [C_F(w), z] \rangle = [C_F(w), z]$$

because $[C_F(v), z] = [C_F(v), z, z] = 1$ by our previous observation and because a and v are conjugate. Thus $F \leq C_G(w)$. □

Now we reach a contradiction – we know that $F \leq H_w$ by (2.6) and therefore $F \leq [H_w, z] \leq F(H_w)$ by $(*)$. But b and w are conjugate, therefore $\pi(F(H_w)) = \pi_b$ whereas F is a π_b'-group by definition. This is a contradiction, so now we know:

If $N_G(P_0) \leq H\ max\ G$ and $H \neq H_b$, then $F_{\pi_b'}(H) = 1$.

Lemma 6.2 (3) implies that b centralises $F(H)$. As $H \neq H_b$ by assumption, this means, by the Infection Theorem (4), that $E(H) \not\leq H_b$. But as we know from Lemma 13.14 (3) and since $H \neq C$, this is only possible if H has a unique component L and L contains a. However, this implies that a centralises $F(H)$ as well, i.e. that z centralises it, and this is impossible by Lemma 4.2 (6) because $H \neq C$.

We obtain that
$$N_G(P_0) \leq H_b.$$
From here we proceed in two steps.

(i) If $P \in \mathrm{Syl}_p(H_b, V)$, then $N_G(P) \leq H_b$. Every z-invariant p-subgroup of $C_G(a)$ is centralised by z.

PROOF. Assume otherwise. We apply Lemma 6.11 (1) and we let $T_0 \in \mathrm{Syl}_2(C_G(P))$ and $c \in C \cap N_G(P) \cap N_G(T_0)$ be such that $b \neq b^c \in Z(T_0)$. Let $e := b^c$ and let $H_e := C_G(e)$. We have that $P_0 \leq P \leq H_e$ and therefore $P_0 = [P_0, z] \leq F(H_e)$ by $(*)$. As $N_G(P_0) \leq H_b$, this means that $H_e \hookrightarrow H_b$. But then Lemma 13.10 yields a contradiction. The last assertion comes directly from Lemma 6.11 (2). □

(ii) As in our initial notation let $v \in C_C(a)$ be an involution such that $\langle v, a \rangle$ is a fours group contained in a 2-component of C and let $w := vz$. Then P_0 is centralised by v.

PROOF. Let $E \in \mathcal{L}_2(C)$ be such that $\langle a, v \rangle \leq E$. Then there exists an odd number $q \geq 11$ such that $\overline{E} \simeq PSL_2(q)$ as on List IV and therefore $\overline{a}$, $\overline{v}$ and $\overline{av}$ are conjugate by some element of order 3 in $\overline{E}$. Thus a, v and av are conjugate in C by some element of odd order. This implies that b and w are commuting involutions that are conjugate in C. The coprime action of $\langle w, z \rangle$ on P_0 yields that

$$P_0 = \langle [C_{P_0}(z), z], [C_{P_0}(v), z], [C_{P_0}(w), z] \rangle = \langle [C_{P_0}(v), z], [C_{P_0}(w), z] \rangle,$$

by Lemma 2.1 (4). We set $P_1 := [C_{P_0}(w), z]$. Then P_1 is centralised by b and by w, hence by $bw = av$, an involution that is conjugate to a. As P_1 is z-invariant and as $P_1 = [P_1, z] \leq C_G(av)$ and $C_G(av)$ is C-conjugate to $C_G(a)$, we know from (i) that P_1 is centralised by z. Thus $P_1 = 1$ by Lemma 2.1 (2). It follows that $P_0 = [C_{P_0}(v), z]$, i.e. v centralises P_0 as stated. □

Statement (ii) and the main hypothesis imply that P_0 is a non-trivial z-invariant p-subgroup of $C_G(v)$. Then (i), applied to $C_G(v)$ instead of $C_G(a)$, yields that $P_0 = [P_0, z] = 1$. This is a contradiction and therefore the theorem is proved.

□

LEMMA 14.5. *Suppose that Hypothesis* 14.1 *holds and that a and b are not conjugate. Let* $v \in \{a,b\}$ *and suppose that* $C = M$ *and* $C_G(v) < H_v$. *Then* U_v *is centralised by* vz.

PROOF. With Theorem 13.23 let p be an odd prime such that H_v is of characteristic p. Then, as $C = M$ and $z \notin Z(H_v)$, Lemma 4.10 implies that $p \notin \pi$. We let $U_0 := [C_{U_v}(z), v]$. If $U_0 \neq 1$, then U_0 is a nilpotent $C_C(v)$-invariant subgroup of C of odd order, and therefore we may apply Lemma 13.12. Then $U_0 = [U_0, v] \leq O(F(C))$. But $p \notin \pi$, so $O_p(C) = 1$ and hence $U_0 = 1$. This implies that $C_{U_v}(z) \leq C_{U_v}(v)$ and thus, with Lemma 2.1 (4), we see that $U_v = [U_v, v] \leq [U_v, z] \cap C_G(vz)$. □

LEMMA 14.6. *Suppose that Hypothesis* 14.1 *holds. Then either* $C_G(a)$ *and* $C_G(b)$ *are both maximal or neither of them is.*

PROOF. This statement is immediate if a and b are conjugate. If a and b are not conjugate, then we first note that Hypothesis 6.6 holds by Lemma 13.7. Suppose that $C_G(b) = H_b$ and assume that there exists some odd prime r such that H_a is of characteristic r. Then Lemma 14.3 implies that $C = M$ and Lemma 14.5 forces U_a to be contained in $C_G(b)$. With the Pushing Down Lemma (3) we deduce that $U_a \leq O_r(H_b)$ and therefore $H_b \looparrowright H_a$, by Lemma 6.9. But also, in particular, we see that $O_r(H_b) \neq 1$ and together with the Infection Theorem (1) this gives that $F^*(H_b) = O_r(H_b)$. This is a contradiction.

If $C_G(a) = H_a$ and r is an odd prime such that H_b is of characteristic r, then we argue as in the previous paragraph if $C = M$. Otherwise $C < M$ and we have a z-minimal subgroup $U \leq O_p(M)$ by hypothesis. If $E(M) \neq 1$, then Hypothesis 7.1 is satisfied. Lemma 13.9 yields that there exists an odd number $q \geq 11$ such that $E(M) \simeq PSL_2(q)$ and hence

$$U \leq \bigcap_{g \in G} M^g$$

by Lemma 7.6. This contradicts Lemma 13.1 (1) and consequently $E(M) = 1$. Thus M has characteristic p. We also recall that $O_r(H_b)$ contains a b-minimal subgroup U_b because $C_G(b) \neq H_b$.

(1) $[U, a] \neq 1$.

PROOF. Otherwise we have that $U \leq F(H_a)$ by the Pushing Down Lemma (3) and hence $H_a \looparrowright M$. This contradicts the Infection Theorem (2) because H_a is not of characteristic p. □

(2) $U_b \leq C$.

PROOF. Set $R := [C_{U_b}(a), b]$ and assume that $R \neq 1$. Then with Lemma 13.1 (1) let $N_G(R) \leq H \max G$. The Pushing Down Lemma (2) yields that $R = [R, b] \leq O_r(H_a)$ and hence H_a and H_b infect H. From (1) we know that $[U, a] \neq 1$ and then Lemma 2.1 (4) implies that $U_0 := [C_U(b), z] \neq 1$. It follows with the Pushing Down Lemma (1) that $U_0 \leq F(H_b)$, in particular $r = p$. Then we deduce that

$$U_0 \leq C_{F(H_b)}(b) \leq C_G(U_b) \leq C_G(R) \leq H,$$

by Lemma 6.7. Applying the Pushing Down Lemma (2), we see that $U_0 \leq O_p(H)$ and therefore $U_0 \leq [O_p(H), z]$. As $|\pi_a| \geq 2$ and $p = r \in \pi_a$, Lemma 6.2 (3) yields that $[O_p(H), z] \leq H_a$ and therefore $U_0 \leq H_a$. But then U_0 is centralised by b and by a, hence by z, and this is a contradiction. Thus $R = 1$ which implies that $U_b = [U_b, b] \leq [U_b, a] \cap C$. □

(3) $U \leq O_p(H_b)$.

PROOF. As $U_b \leq C \leq M$ by (2) and this subgroup is $C_M(b)$-invariant, Lemma 13.12 yields that $U_b = [U_b, b] \leq O_p(M)$. Then $U_b \leq C_{O_p(M)}(z) \leq C_G(U)$ by Lemma 6.7 and therefore $U \leq C_G(U_b) \leq H_b$ by Lemma 6.9. The Pushing Down Lemma (3) yields that $U \leq O_p(H_b)$. □

(4) $[C_U(a), z] \neq 1$.

PROOF. As a and b are supposed to be not conjugate, one of the cases from List IV holds by Lemma 13.5 (2) and therefore there exists a 2-component E of C such that $a \in E$. Moreover there exists a fours group A_0 in E containing a and such that the involutions in A_0 are all conjugate in C. As U is C-invariant and

$$U = [U, z] = \langle [C_U(a_0), z] \mid a_0 \in A_0^{\#} \rangle,$$

by Lemma 2.1 (4), we deduce that for all $a_0 \in A_0^{\#}$, the commutator $[C_U(a_0), z]$ is non-trivial. In particular $[C_U(a), z] \neq 1$ as stated. □

Let $X := [C_U(a), z]$. Then $1 \neq X \leq U \leq O_p(H_b)$ by (2) and (3) and with Lemma 13.1 (1) we let $N_G(X) \leq H \max G$. Then M and H_b infect H and we also see, with the Pushing Down Lemma (2), that $X \leq O_p(H_a)$ and hence $H_a \looparrowright H$. Let $F := F_{\pi'_a}(H)$. As $p \in \pi_a$, we see that F is a p'-group and the Infection Theorem (1) forces F to be inverted by a, b and z. Consequently $F = 1$. Moreover $C_G(V) \leq H$ and therefore Lemma 13.14 (3) implies that $E(H) \leq H_b \cap H_a$ or that H has a component that contains a. In the latter case, we see from the type of this component that $E(H)$ contains a fours group A_0 such that its involutions are conjugate within $E(H)$ and such that $a \in A_0$. Then the fact that U is $E(H)$-invariant and that $X \leq H$ forces that for all $a_0 \in A_0^{\#}$, the subgroup $[C_U(a_0), z]$ is contained in H. With Lemma 2.1 (4) it follows that $U \leq H$ and then $U \leq O_p(H)$ by the Pushing Down Lemma (3). Lemma 6.9 yields that $H \looparrowright M$. But $M \looparrowright H$ and therefore the Infection Theorem (3) gives that $F^*(H) = O_p(H)$ (whether it coincides with M or not does not matter). Then H_a infects a maximal subgroup of characteristic p and this is impossible by the Infection Theorem (2). This last contradiction finishes the case where $C_G(a)$ is a maximal subgroup of G and $C_G(b)$ is not. □

LEMMA 14.7. *Suppose that Hypothesis* 14.1 *holds and that* $C_G(b)$ *is a maximal subgroup of* G. *Then* $O(C) = 1$ *(and therefore the* F^*-*Structure Theorem describes* $F^*(C)$).

PROOF. From Lemma 14.3 and Theorem 14.4 we know that $C = M$ and $O(F(C)) \cap H_b = 1$. If $r_2(G) = 2$, then Lemma 14.2 implies that H_b has odd prime characteristic, which is a contradiction. Hence $r_2(G) = 3$, more precisely

V is contained in an elementary abelian subgroup of order 8. Let $d \in C$ be an involution distinct from a and b, but commuting with them and such that a, d and ad are conjugate in C. Choose $c \in C$ to be such that $a^c = d$. Now our hypothesis implies that b inverts $O(F(C))$. Then also a inverts it and $d = a^c$ inverts $O(F(C))^c = O(F(C))$, similarly ad inverts it by conjugacy whereas it also centralises it. Since $O(F(C))$ has odd order, this yields that $O(F(C)) = 1$. But then $F(O(C)) = 1$ whence $O(C) = 1$. □

LEMMA 14.8. *Suppose that Hypothesis 14.1 holds and that $C_G(b)$ is a maximal subgroup of G. Let $p \in \pi(F(H_b))$ be an odd prime such that $O_p(H_b) \not\leq C$ and let $P_0 := [O_p(H_b), z]$. Let $P \in Syl_p(H_b, V)$. Then $N_G(P_0) \leq H_b$ or $N_G(P) \leq H_b$.*

PROOF. First we apply Lemma 14.7 to deduce that $O(C) = 1$. We assume that $N_G(P_0) \leq H \max G$ with $H \neq H_b$ and that $N_G(P) \not\leq H_b$, aiming for a contradiction. Then $H_b \looparrowright H$. We recall that $|\pi_b| \geq 2$ because π_b contains 2 and p. Therefore Lemma 6.2 (3) and the Infection Theorem (1) yield that $F_{\pi_b}(H)$ is centralised by b and that $F_{\pi_b'}(H)$ intersects H_b trivially and is therefore inverted by b. In particular $[H, b]$ centralises $F(H)$.

(1) $E(H) \leq H_b$.

PROOF. As $C_G(V) \leq H$, Lemma 13.14 is applicable and gives that $E(H) \leq H_b$ or that H possesses a component that contains a. In the latter case it follows that $F(H)$ is centralised by a. Thus z centralises $E(H)$ and $F_{\pi_b}(H)$, and it inverts $F_{\pi_b'}(H)$. We deduce that $[H, z]$ centralises $F^*(H)$, therefore $[H, z] \leq Z(F(H))$ and hence

$$P_0 = [P_0, z] \leq O_p(H) \leq C_G(a).$$

But then P_0 is centralised by b and by a, hence by z, and this is a contradiction. Thus $E(H) \leq H_b$ as stated. □

(2) Let $F := F_{\pi_b'}(H)$. Then $C_F(z) = 1$, but $F \neq 1$, and moreover $H_a \looparrowright H$.

PROOF. Step (1) implies that $[H, b] \leq C_H(F^*(H)) = Z(F(H))$. Since $C_C(a) = C_C(b) \leq N_G(P_0) \leq H$, the group F is a $C_C(b)$-invariant 2'-subgroup of H. Thus $C_F(z)$ is $C_C(b)$-invariant and inverted by b. We recall that $O(C) = 1$ and apply Lemma 13.12 to see that $C_F(z) = 1$ as stated. In particular F is inverted by b and z. Then $F \leq C_G(a)$ and we know from Lemma 14.6 that $C_G(a) = H_a$, so F is a $C_{H_a}(z)$-invariant nilpotent subgroup of H_a. The Pushing Down Lemma (1) implies that $F = [F, z] \leq F(H_a)$. Since $E(H) \leq C \cap H_b$ by Lemma 4.2 (5) and by (1) (and hence $E(H)$ is centralised by a) and since $H \neq H_b$ by assumption, the Infection Theorem (4) tells us that $F \neq 1$. Then $N_G(F) = H$ by Corollary 4.8 and it follows that $H_a \looparrowright H$. □

(3) a and b are not conjugate.

PROOF. If a and b are conjugate, then $\pi_a = \pi_b$ and therefore (2) and the Infection Theorem (1) imply that $F = F_{\pi_a'}(H)$ is inverted by a. This is only possible if $F = 1$, but this contradicts (2). □

(4) $P_0 = O_p(H_b)$.

PROOF. We use (1) and also our assumption that $N_G(P) \not\leq H_b$. Let $T_0 \in \mathrm{Syl}_2(C_G(P))$. We may suppose that T_0 is z-invariant by Lemma 3.11, applied to $C_G(P)\langle z \rangle$. With Lemma 6.11 (1) we let $c \in N_G(P) \cap N_G(T_0) \cap C$ be such that $v := b^c$ is an involution in $Z(T_0)$ distinct from b. Then $P = P^c \leq H_b^c =: H_v$ which means that P is centralised by b and by v. Let $X := C_{O_p(H_b)}(z)$. Then X is a $C_C(a)$-invariant p-subgroup of C that is centralised by b and by v. We need to recall that a and b are not conjugate by (3) and that, therefore, C possesses a simple component E by Lemma 13.5 (2) and since $O(C) = 1$. More precisely there exists an odd number $q \geq 11$ such that $E \simeq PSL_2(q)$. Then $[X, C_E(a)] \leq X \cap E$ and we note that $O(C_E(a))$ is cyclic and inverted by v (because $C_E(a)$ is an involution centraliser in $PSL_2(q)$). As X is centralised by v, it follows that

$$[X, C_E(a), v] \leq [X, v] = 1 \text{ and } [v, X, C_E(a)] = 1,$$

so the Three Subgroups Lemma implies that $[C_E(a), v, X] = 1$. We recall that $O(C_E(a))$ is inverted by v and therefore $[O(C_E(a)), X] \leq [C_E(a), v, X] = 1$. As X is a p-group, this shows that $[C_E(a), X] = 1$ and hence X cannot induce non-trivial inner automorphisms on E because v centralises X. But also, Remark 10.2 says that X does not induce a non-trivial field automorphism on E, so we deduce that X centralises E. If C possesses a second component E_1, then E_1 and $O_2(C)$ (being subgroups of $C_C(a)$) normalise X and therefore $[O_2(C)E_1, X] = 1$. This forces

$$X \leq C_C(F^*(C)) = Z(F(C)) \leq O_2(C)$$

and hence $X = 1$. We conclude that $P_0 = O_p(H_b)$. $\square$

From (4) and Corollary 4.8 it follows that

$$H_b = N_G(O_p(H_b)) = N_G(P_0) \leq H$$

and consequently $H_b = H$, contrary to our assumption. Thus the lemma is proved. $\square$

All previous statements together with some additional work yield the first main result of this section.

THEOREM 14.9. *Suppose that Hypothesis 14.1 holds and that $C_G(a)$ is a maximal subgroup of G. Then a and b are not conjugate.*

PROOF. We assume, by way of contradiction, that a and b are conjugate. Then, by Lemma 13.5 (2), we know that one of the cases from List I, II or III holds and moreover our hypothesis implies that $C_G(b) = H_b$. With Lemma 14.2 and Theorem D we know that $r_2(G) = 3$, and Lemma 14.7 yields that $O(C) = 1$.

As $z \notin Z(H_a)$, we choose, with Lemma 4.2 (6), an odd prime p such that $P_0 := [O_p(H_a), z] \neq 1$. With Lemma 13.5 (8) let $P \in \mathrm{Syl}_p(H_a, V)$. Then Lemma 6.11 (1) implies that $N_G(P) \not\leq H_a$ and then, with Lemma 14.8, that $N_G(P_0) \leq H_a$.

Next we look at $X := C_{O_p(H_a)}(z)$ and $Y := C_{O_p(H_b)}(z)$. We assume in the following that these subgroups are non-trivial. They are conjugate by our assumption that a and b are conjugate. We also note that X and Y normalise each other whence XY is a p-group.

(*) $X = Y$.

PROOF. If $F^*(C)$ is as on List II, then let E be the unique component of C. Then $X \cap E$ is a $C_E(a)$-invariant p-subgroup of $C_E(a)$ and hence centralised by $C_E(a)$ and $Y \cap E$ is a $C_E(a)$-invariant p-subgroup of $C_E(a)$, thus also centralised by $C_E(a)$. It follows that $X \cap E = Y \cap E$. Moreover we know that the outer automorphism group of E has cyclic Sylow p-subgroups, so the groups of outer automorphisms that X and Y induce on E coincide. We deduce that $XC_C(E) = YC_C(E)$.

Now suppose that $F^*(C)$ is as on List I or III and hence that a is diagonal. As X and Y are $C_C(a)$-invariant p-groups, they cannot be contained in $O_2(C)$ or in a component of C that centralises a. For every component of C, the outer automorphism group has cyclic Sylow p-subgroups and as X and Y normalise every component of C, we see that X and Y induce the same outer automorphisms on every component of C. They also induce the same inner automorphisms and therefore $XC_C(E(C)) = YC_C(E(C))$ as in the previous case.

Hence if we let $x \in X$, then there exists some $y \in Y$ such that x and y induce exactly the same automorphism on $E(C)$ whence $x^{-1}y \in XY$ is a p-element in C that centralises $E(C)$. As XY is $C_C(a)$-invariant, we also see that XY centralises $O_2(C)$ and hence $x^{-1}y \in C$ is a p-element that centralises $F^*(C)$. But $C_C(F^*(C)) = Z(O_2(C))$, so we see that $x = y$ and it follows that $X = Y$. □

Let $N_G(X) \le H \max G$. Then $H_a \hookrightarrow H$ and by (*) we also know that $H_b \hookrightarrow H$. Now we recall that a and b are conjugate by assumption and that, therefore, we have that $\pi_a = \pi_b$. Then with Lemma 6.2 (3) and the Infection Theorem (1) it follows that a and b invert $F_{\pi'_a}(H)$ and centralise $F_{\pi_a}(H)$. Thus z centralises $F^*(H)$ and Lemma 4.2 (6) forces $H = C$. So H_a and H_b infect C. Moreover $F_{p'}(H_a)$ and $F_{p'}(H_b)$ are contained in $C_G(X)$ and hence in C, so in particular $X \ne O_p(H_a)$ because $z \notin C_{H_a}(F(H_a))$. Let $X_1 := N_{O_p(H_a)}(X)$. Then $X < X_1 \le H = C$ and thus $X_1 \le C_{O_p(H_a)}(z) = X$, which is a contradiction. □

LEMMA 14.10. *Suppose that Hypothesis 14.1 holds and that $C_G(b)$ is a maximal subgroup of G. Then there exist a prime p and some $P \in Syl_p(H_b, V)$ such that $[O_p(H_b), z] \ne 1$ and $N_G(P) \le H_b$.*

PROOF. Assume otherwise. Hence for all $p \in \pi_b$ and for all $P \in \mathrm{Syl}_p(H_b, V)$ we have that, if $[O_p(H_b), z] \ne 1$, then $N_G(P) \not\le H_b$. With Lemma 4.2 (6) we choose $p \in \pi_b$ such that $P_0 := [O_p(H_b), z] \ne 1$. By Lemma 14.8 and our assumption we have that $N_G(P_0) \le H_b$. Also, with Lemma 6.11 (1), we find an element $c \in C \cap N_G(P)$ such that $d := b^c$ is an involution distinct from b, commuting with b and z and centralising P. Let $H_d := H_b^c$. Then $H_d = C_G(d)$ by conjugacy and $O_p(H_b) \le P \le H_d$. Therefore $X := C_{O_p(H_b)}(z)$ is centralised by d. Looking at C, we first recall that $O(C) = 1$ by Lemma 14.7 and then that $F^*(C)$ is as in one of the cases from List IV, because a and b are not conjugate by Theorem 14.9 and by Lemma 13.5 (2). We denote the simple component of C by E and note that $C_E(a)(= C_E(b))$ and X normalise each other. Moreover we recall that d is conjugate to b and commutes with it, therefore dz is conjugate to a and commutes with it and it follows that dz inverts $O(C_E(a))$. At the same time d and z centralise X and hence dz does, so we have that $[dz, X, C_E(a)] = 1$. We use this information to show that X centralises E. First $[X, C_E(a)] \leq X \cap E \leq C_G(dz)$ and therefore $[X, C_E(a), dz] = 1$. Then $[C_E(a), dz, X] = 1$ with the Three Subgroups Lemma. As dz inverts $O(C_E(a))$ and X is a p-group, we see first that X centralises $O(C_E(a))$ and then that X centralises all of $C_E(a)$. Thus the elements from $X^{\#}$ cannot induce (non-trivial) inner automorphisms on E, and by Remark 10.2 they also cannot induce outer automorphisms. It follows that $[E, X] = 1$ as stated. As $O_2(C) \leq C_C(a) = C_C(b)$ by choice of a, we also have that X centralises $O_2(C)$. If C has a second component, then this component centralises a and z, hence b, therefore normalises X and is X-invariant, so it centralises X. All this forces

$$X \leq C_C(F^*(C)) = Z(F^*(C))$$

which is a 2-group, so $X = 1$.

From our hypothesis it follows that we can argue in this way for all odd primes in π_b and we obtain, for all $r \in \pi_b$, that $O_r(H_b)$ is either centralised by z (which includes the case $r = 2$ by Lemma 4.2 (5)) or inverted by it. Hence $[H_b, z] \leq Z(F(H_b))$. Going back to our prime p, the subgroup P_0 and the involution d, conjugacy yields that $P_0 = [P_0, z] \leq [H_d, z] \leq Z(F(H_d))$ and therefore $H_d \looparrowright H_b$. But b and d are conjugate, so this is a contradiction to Lemma 13.10. $\square$

THEOREM 14.11. *Suppose that Hypothesis 14.1 holds. Then $C_G(a)$ is not a maximal subgroup of G.*

PROOF. Assume otherwise. Then Lemma 14.6 yields that $C_G(b)$ is a maximal subgroup as well and we know from Theorem 14.9 that a and b are not conjugate. With Lemma 14.10 we choose a prime p and a V-invariant Sylow p-subgroup P of H_b such that $[O_p(H_b), z] \neq 1$ and $N_G(P) \leq H_b$. With Lemma 6.11 (2) we see that for this prime p, every $\langle z \rangle$-invariant p-subgroup of $C_G(a)$ is centralised by z. Now we recall that a and b are not conjugate and that, therefore, by Lemmas 13.5 (2) and 14.7, we have a simple component E in C. Then $a \in E$ by hypothesis, we choose a fours group in E containing a and we denote its involutions by a, v and av. Then a, v and av are C-conjugate. As every $\langle z \rangle$-invariant p-subgroup of $C_G(a)$ is centralised by z, conjugacy implies that also every $\langle z \rangle$-invariant p-subgroup of $C_G(v)$ is centralised by z. Moreover we note that b and $w := vz$ are conjugate and that $\langle w, z \rangle$ acts coprimely on $P_0 := [O_p(H_b), z]$. Lemma 2.1 (4) yields that

$$P_0 = [P_0, z] = \langle [C_{P_0}(z), z], [C_{P_0}(v), z], [C_{P_0}(w), z] \rangle = \langle [C_{P_0}(v), z], [C_{P_0}(w), z] \rangle.$$

As $[C_{P_0}(v), z]$ is a z-invariant p-subgroup of $C_G(v)$, we have that $[C_{P_0}(v), z, z] = 1$ and hence $[C_{P_0}(v), z] = 1$. This implies that $P_0 = [C_{P_0}(w), z]$ and therefore P_0 is centralised by w and by b. It follows that bw centralises P_0. But $bw = azvz = av$ is conjugate to a, so conjugacy yields that the z-invariant p-subgroup P_0 of $C_G(av)$ must be centralised by z. Thus $P_0 = [P_0, z] = 1$, which is a contradiction. $\square$

HYPOTHESIS 14.12. *In addition to Hypothesis 10.1 we suppose the following:*
- $a \in C$ *is an involution distinct from z and chosen as in Lemma* 13.4.
- $V := \langle a, z \rangle$.
- *M is a maximal subgroup containing C such that, if possible, there exists an odd prime r such that $O_r(M) \neq 1 = C_{O_r(M)}(z)$. If $C < M$, then let $U \leq O(F(M))$ be a z-minimal subgroup.*
- *For all $v \in \{a, b\}$ we let H_v be a maximal subgroup of G such that $C_G(v) < H_v$ and let p be an odd prime such that $F^*(H_v) = O_p(H_v)$. If possible, we suppose that v inverts $F^*(H_v)$. Let $U_v \leq O_p(H_v)$ denote a v-minimal subgroup.*

We note that Hypothesis 14.12 implies Hypothesis 13.6 and therefore, by Lemma 13.7, also Hypothesis 6.6.

LEMMA 14.13. *Suppose that Hypothesis* 14.12 *holds. Let r be an odd prime and let $H < G$ be such that $C_G(V) \leq H$. Then $\widehat{H} := H/O(H)$ has a unique maximal $\widehat{C_G(V)}$-invariant r-subgroup.*

PROOF. In C we can choose a Sylow 2-subgroup S such that S contains a Sylow 2-subgroup S_0 of $C_G(V)$. Let $V \leq S_0 \leq S_1 \in \mathrm{Syl}_2(H)$. Then every member of $\mathcal{M}_H(C_G(V), r)$ is S_0-invariant and we know, from the structure of $F^*(\overline{C})$ (or explicitly from (i) in the proof of Lemma 13.4), that $S_0 = S$ or that S_0 has index 2 in S. In particular $S_0 = S_1$ or S_0 has index 2 in S_1. Let $\widehat{Y} \in \mathcal{M}_{\widehat{H}}(\widehat{C_G(V)}, r)$. As $O_2(\widehat{H}) \leq \widehat{S_1}$, we see that $[C_{O_2(\widehat{H})}(\widehat{V}), \widehat{Y}] \leq O_2(\widehat{H}) \cap \widehat{Y} = 1$ whence $\widehat{Y}$ centralises a subgroup of index at most 2 of $O_2(\widehat{H})$. It follows that $[O_2(\widehat{H}), \widehat{Y}] = 1$.

If $\widehat{H}$ has characteristic 2, then $\widehat{Y} = 1$ and our claim is proved. Thus let us suppose that $\widehat{H}$ has a component. Then $\mathcal{L}_2(H) \neq \varnothing$ and for all $L \in \mathcal{L}_2(H)$, the subgroup $L_0 := O^\infty(C_L(z))$ lies in C. Moreover $\overline{L_0}$ is not soluble, therefore induces inner automorphisms on $F^*(\overline{C})$ and is hence contained in it. Thus, going through the lists of the F*-Structure Theorem, the possibilities for $C_{\widehat{L}}(\widehat{V})$ are as follows:

- $C_{\widehat{L}}(\widehat{V}) = \widehat{L}$ or
- $C_{\widehat{L}}(\widehat{V})$ is cyclic or dihedral or
- there exists an odd prime power $q \geq 5$ such that $C_{\widehat{L}}(\widehat{V})$ has two components of type $SL_2(q)$, with central involutions $\widehat{a}$ and $\widehat{b}$.

We deduce that, as $\widehat{Y}$ is $C_{E(\widehat{H})}(\widehat{V})$-invariant, it induces inner automorphisms on every component of $\widehat{H}$. Thus $\widehat{Y} \leq F^*(\widehat{H})$. But then, from the possibilities above and the $C_{F^*(\widehat{H})}(\widehat{V})$-invariance of $\widehat{Y}$, it follows that $\widehat{Y} \leq C_{F^*(\widehat{H})}(\widehat{V})$. We already know that $\widehat{Y}$ centralises $O_2(\widehat{H})$. For each component $\widehat{L}$, going through the possibilities for $C_{\widehat{L}}(\widehat{V})$ yields the following:

- if $C_{\widehat{L}}(\widehat{V}) = \widehat{L}$, then $\widehat{Y} \cap \widehat{L} = 1$.
- if $C_{\widehat{L}}(\widehat{V})$ is cyclic or dihedral, then $\widehat{Y} \cap \widehat{L}$ is contained in it.
- if $C_{\widehat{L}}(\widehat{V})$ has two components of type $SL_2(q)$ with central involutions $\widehat{a}$ and $\widehat{b}$, then $\widehat{Y} \cap \widehat{L}$ centralises $C_{\widehat{L}}(\widehat{V})$ because $\widehat{Y}$ has odd order.

If $\widehat{Y} \in \mathsf{M}^*_{\widehat{H}}(\widehat{C_G(V)}, r)$, then we conclude that either $\widehat{Y} \cap \widehat{L}$ is contained in $C_{\widehat{L}}(\widehat{V})$, that it even is a maximal $C_{\widehat{L}}(\widehat{V})$-invariant r-subgroup there and that it therefore coincides with $O_r(C_{\widehat{L}}(\widehat{V}))$, or that $F^*(\overline{C})$ is as on List II and $\widehat{Y} \cap \widehat{L}$ centralises $C_{\widehat{L}}(\widehat{V})$. In this case $\widehat{Y} \cap \widehat{L}$ is also a maximal $C_{\widehat{H}}(\widehat{V})$-invariant r-subgroup of $\widehat{L}$.

We deduce that $\widehat{Y} \cap \widehat{L}$ is the unique maximal $C_{\widehat{L}}(\widehat{V})$-invariant r-subgroup of $\widehat{L}$.

Now let $\widehat{Y_1}, \widehat{Y_2} \in \mathsf{M}^*_{\widehat{H}}(\widehat{C_G(V)}, r)$. Then $\widehat{Y_1}$ and $\widehat{Y_2}$ are contained in $F^*(\widehat{H})$ and even in $C_{F^*(\widehat{H})}(\widehat{V}) = C_{O_2(\widehat{H})E(\widehat{H})}(\widehat{V})$, as we deduced in the previous paragraph.

But r is odd and hence $\widehat{Y_1}, \widehat{Y_2} \leq C_{E(\widehat{H})}(\widehat{V})$. Conversely, we know for every component $\widehat{L}$ of $\widehat{H}$ that $\widehat{Y_1} \cap \widehat{L}$ and $\widehat{Y_2} \cap \widehat{L}$ coincide with the unique maximal $C_{\widehat{L}}(\widehat{V})$-invariant r-subgroup $\widehat{R_L}$ of $\widehat{L}$. Thus

$$\widehat{Y_1} = \widehat{Y_1} \cap E(\widehat{H}) = \prod_{L \in \mathcal{L}_2(H)} \widehat{R_L} = \widehat{Y_2} \cap E(\widehat{H}) = \widehat{Y_2}.$$

□

THEOREM 14.14. *Hypothesis 14.12 does not hold.*

PROOF. Assume otherwise, with the notation from Hypothesis 14.12. First we show that (H_a, H_b) is a V-special primitive pair of characteristic p as in Definition 2.25. As H_a and H_b are primitive (by Corollary 4.8) and of characteristic p, we only need to verify two properties:
- $V \leq Z^*(H_a) \cap Z^*(H_b)$ holds because of Lemma 13.5 (1).
- $C_G(V) \leq H_a \cap H_b$ holds because $a, b \in V$.

Therefore (H_a, H_b) is in fact a V-special primitive pair of characteristic p of G. Then Lemma 14.13 implies that Theorem 2.26 is applicable and it yields that $O_p(H_a) \cap H_b = 1 = O_p(H_b) \cap H_a$. In particular $O_p(H_a) \cap O_p(H_b) = 1$ and $O_p(H_a)$ is inverted by b. Therefore $O_p(H_a)$ is abelian and this implies that U_a is elementary abelian and that a inverts it. It follows that U_a is inverted by a and by b and therefore $U_a \leq C$. Then Lemma 13.12 yields that $U_a = [U_a, a] \leq O_p(M)$. From Lemma 4.2 (6) we know that $O_p(H_a)$ is not centralised by z, but we just saw that $p \in \pi(F(M))$. Therefore $C \neq M$ by Lemma 4.10. In particular, we have our z-minimal subgroup U and we conclude that $U_a \leq C_{O_p(M)}(z) \leq C_G(U)$ with Lemma 6.7. Then Lemma 6.9 implies that $U \leq C_G(U_a) \leq H_a$ and the Pushing Down Lemma (3) forces $U \leq O_p(H_a)$. We conclude that, since $O_p(H_a) \cap H_b = 1$, we have that $U \leq O_p(H_a)$. Now we recall that also $O_p(H_b) \cap H_a = 1$, so a inverts $O_p(H_b)$ and by symmetry between a and b we deduce that $U \leq O_p(H_b)$ whereas $O_p(H_a) \cap O_p(H_b) = 1$. This is impossible. □

CHAPTER 15

The Final Contradiction and the Z*-Theorem for $\mathcal{K}_2$-Groups

In the previous chapter we concluded our analysis of maximal subgroups of G containing an involution centraliser. Without too much effort, we can now reach a contradiction under the hypothesis that, with all the notation in our minimal counter-example G, the components of $\overline{C}$ are known quasi-simple groups:

THEOREM 15.1. *Hypothesis 10.1 does not hold.*

PROOF. Assume otherwise. Then in particular we have Hypothesis 4.1 with all the notation included there. The F*-Structure Theorem is applicable and yields that we know precisely what the possibilities for the shape of $F^*(\overline{C})$ are. Next we choose an involution $a \in C$ as in Lemma 13.4. Then Theorem 13.23 implies that either $C_G(a)$ is a maximal subgroup of G itself or it is properly contained in a maximal subgroup H_a of G of odd prime characteristic. With this choice of the involution a and a suitable choice for a maximal subgroup M containing C, we can set notation such that Hypothesis 14.1 holds. Therefore we may apply Theorem 14.11 and we deduce that $C_G(a)$ is not a maximal subgroup. Thus we know that $C_G(a)$ is properly contained in a maximal subgroup H_a of G and that there exists an odd prime p such that $F^*(H_a) = O_p(H_a)$. If possible, we choose H_a such that a inverts $O_p(H_a)$. Lemma 14.6 yields that $C_G(az)$ is properly contained in a maximal subgroup H_{az} of G of characteristic p and we choose H_{az} such that az inverts $O_p(H_{az})$, if this is possible. Then Hypothesis 14.12 holds and Theorem 14.14 provides a contradiction. □

If we want to state an independent result, then we must somehow capture the knowledge about simple groups involved in centralisers of isolated involutions that we incorporated in Hypothesis 10.1 in our minimal counter-example. We recall that, by the Classification of the Finite Simple Groups (as stated for example in [Wil09]), every finite simple group is isomorphic to one of the following:

- a cyclic group of prime order,
- an alternating group A_n for some $n \geq 5$,
- a simple group of Lie type,
- a sporadic simple group.

The Z*-Theorem for $\mathcal{K}_2$-groups is an attempt to capture, in a suitable hypothesis, the fact that we do not really need information about all proper subgroups of G, but only about centralisers of isolated involutions.

DEFINITION 15.2. We say that a finite group X is a $\mathcal{K}_2$-group if and only if for every isolated involution $t \in X$ and every subgroup H of X containing t, all simple sections (i.e. factor groups of subgroups) in $C_H(t)/O(C_H(t))$ are known simple groups. By a "known simple group" we mean a group from the list in the Classification of Finite Simple Groups.

We note that the definition implies that subgroups and factor groups of $\mathcal{K}_2$-groups are themselves $\mathcal{K}_2$-groups. In the literature, the term $\mathcal{K}_2$-group is sometimes used to describe simple groups where in all 2-local subgroups, all simple sections are known simple groups.

PROOF OF THE Z*-THEOREM FOR $\mathcal{K}_2$-GROUPS.

Assume that this result is false and choose G to be a minimal counter-example. Let then $z \in G$ be an isolated involution such that $z \notin Z^*(G)$. In order to establish Hypothesis 4.1 for G, we look at an isolated involution t of G and we suppose that $t \in H < G$. Then our hypothesis implies that H is a $\mathcal{K}_2$-group. As t is isolated in H, the minimal choice of G yields that the theorem holds in H. This means that $t \in Z^*(H)$. Next let $N \trianglelefteq G$ be such that $t \notin N$ and set $\widehat{G} := G/N$. Then $\widehat{t}$ is isolated in $\widehat{G}$ by Lemma 3.1 (7). Hence if $N \neq 1$, then $\widehat{G}$ is a proper factor group of G and the minimality of G yields that $\widehat{t} \in Z^*(\widehat{G})$. Therefore Hypothesis 4.1 holds for G and, with our additional $\mathcal{K}_2$-group hypothesis, we even have that Hypothesis 10.1 holds. This contradicts Theorem 15.1. □

Bibliography

[Asc00]   M. Aschbacher, *Finite Group Theory*, Cambridge University Press, 2000. MR1777008 (2001c:20001)

[Ben81]   H. Bender, Finite Groups with Dihedral Sylow 2-Subgroups, *J. Algebra* **70** (1981), 216–228.

[BG81]   H. Bender and G. Glauberman, Characters of Finite Groups with Dihedral Sylow 2-Subgroups, *J. Algebra* **70** (1981), 200–215.

[BG94]   H. Bender and G. Glauberman, *Local Analysis for the Odd Order Theorem*, Cambridge University Press, 1994. MR1311244 (96h:20036)

[Bro83]   M. Broué, *La Z_p^*-conjecture de Glauberman (Séminaire sur les Groupes Finis, tome 1)*, Publications de l'Univ. Paris VII (1983).

[CCN103] J. Conway, R. Curtis, S. Norton, R. Parker, and R. Wilson, ATLAS *of Finite Groups*, Oxford University Press, 2003.

[Cra11]   D. A. Craven, *The Theory of Fusion Systems*, Cambridge University Press, 2011. MR2808319 (2012f:20028)

[FT63]   W. Feit and J.G. Thompson, *Solvability of groups of odd order*, Pacific J. Math. **13** (1963), 775–1029. MR0166261 (29:3538)

[Fis64]   B. Fischer, *Distributive Quasigruppen endlicher Ordnung*, Math. Z. **83** (1964), 267–303. MR0160845 (28:4055)

[Gla66a] G. Glauberman, *Central Elements in Core-Free Groups*, J. Algebra **4** (1966), 403–420. MR0202822 (34:2681)

[Gla66b] ———, *On the Automorphism Group of a Finite Group Having No Non-Identity Normal Subgroups of Odd Order*, Math. Z. **93** (1966), 154–160. MR0194503 (33:2713)

[Gla68a] ———, *A characteristic subgroup of a p-stable group*, Canadian J. Math. **20** (1968), 1101–1135. MR0230807 (37:6365)

[Gla68b] ———, *On Loops of Odd Order II*, J. Algebra **8** (1968), 393–414. MR0222198 (36:5250)

[Gla72]   ———, *Fixed Point Subgroups that Contain Centralizers of Involutions*, Math. Z. **124** (1972), 353–360. MR0291269 (45:363)

[Gla74]   ———, *On groups with a quaternion Sylow 2-subgroup*, Illinois J. Math. **118** (1974), 60–65. MR0332969 (48:11294)

[Gla76]   ———, *On solvable signalizer functors in finite groups*, Proc. London Math. Soc **33** (1976), 1–27. MR0417284 (54:5341)

[Gol72]   D. M. Goldschmidt, *Weakly Embedded 2-Local Subgroups of Finite Groups*, J. Algebra **21** (1972), 341–351. MR0323896 (48:2249)

[Gol75]   ———, *Strongly Closed 2-Subgroups of Finite Groups*, Ann. Math.(II) **102, No.3** (1975), 475–489. MR0393223 (52:14033)

[Gor82]   D. Gorenstein, *Finite Simple Groups*, Plenum Press, 1982.

[GLS96]   D. Gorenstein, R. Lyons, and R. Solomon, *The Classification of the Finite Simple Groups, Number 2*, American Mathematical Society, 1996. MR1358135 (96h:20032)

[GLS98]   ———, *The Classification of the Finite Simple Groups, Number 3*, American Mathematical Society, 1998. MR1490581 (98j:20011)

[GR93]   R. Guralnick and G. Robinson, *On extensions of the Baer-Suzuki Theorem*, Israel J. Math **82** (1993), 281–297. MR1239051 (94i:20034)

[KS04]   H. Kurzweil and B. Stellmacher, *The Theory of Finite Groups*, Springer, 2004. MR2014408 (2004h:20001)

[Row81]   P. Rowley, *3-locally central elements in finite groups*, Proc. London Math. Soc. **43** (1981), 357–384. MR628282 (82m:20013)

[Wal08] R. Waldecker, *A theorem about coprime action*, J. Algebra **320** (2008), 2027–2030. MR2437641 (2009h:20025)
[Wal09] _____, *Isolated involutions whose centraliser is soluble*, J. Algebra **321** (2009), 1561–1592. MR2494410 (2010a:20053)
[Wal11] _____, *Special primitive pairs in finite groups*, Archiv der Mathematik **97, No. 1** (2011), 11–16. MR2820583
[Wil09] R. A. Wilson, *The Finite Simple Groups*, Springer, 2009.

Index

2-balanced, 29
2-component, 8
$2A_n$, 8
$2J_2$, 8
$3A_7$, 8
$3PSL_2(9)$, 8
A_n, 8
C_n, 8
$F^*(\overline{C})$ is as on List ..., 93
$F_\pi(X)$, 7
$H \max X$, 7
$H_1 \looparrowright H_2$, 37
H_1 infects H_2, 37
$I_Y(t)$, 7
$L(X)$, 8
$O(X)$, 7
$O^p(X)$, 7
$O^\infty(X)$, 7
$O_{2'}(G)$, 1
$O_{\pi',E}(X)$, 8
$O_{\pi',F^*}(X)$, 8
$O_{\pi',F}(X)$, 8
$O_{\pi',\pi}(X)$, 8
$O_{\pi'}(X)$, 7
Y^U, 7
$ZJ(X)$, 7
$Z^*(X)$, 7
$Z_\pi^*(X)$, 7
$\mathcal{K}_2$-group, 146
$\mathcal{L}_2(X)$, 8
$\Theta(a)$, 29
$\alpha(a)$, 29
$\gamma(a)$, 29
$\mathcal{U}_X(A,\pi)$, 8
$\mathcal{U}_X^*(A,\pi)$, 8
π-component, 8
$\mathrm{Syl}_p(H,V)$, 21
$\mathrm{Syl}_p(H,z)$, 21
$\mathcal{S}_n$, 8
$a \circ b$, 18
$a + b$, 19
n_p, 7
p-nilpotent, 8

p-perfect, 7
p-rank, 8
$r(X)$, 8
$r_p(X)$, 8
t-minimal subgroup, 40
x^U, 7

alternating group, 8

balanced, 29
Bender Method, 37

central product, 8
centraliser closed, 8
characteristic p, 7
complete, 30
completion, 30
component, 8
core, 7
core-separated, 32

diagonal involution, 8

Glauberman's Z*-Theorem, 1
Gorenstein-Walter Theorem, 12

Hypothesis 10.1, 75
Hypothesis 13.15, 102
Hypothesis 13.6, 98
Hypothesis 14.1, 123
Hypothesis 14.12, 142
Hypothesis 4.1, 23
Hypothesis 6.6, 40
Hypothesis 7.1, 45
Hypothesis 8.1, 55
Hypothesis 8.12, 63

infection, 37
Infection Theorem, 38
involution, 1
isolated in, 7
isolated with respect to, 1

L-Balance, 69
List (a), 81
List (b), 81

List I, 91
List II, 91
List III, 91
List IV, 91

max, 7

nilpotent action, 8
normal p-complement, 8

odd prime characteristic, 8
operation $+$ on K, 19
operation $\circ$ on K, 18

prime characteristic, 8
primitive subgroup, 7
Pushing Down Lemma, 40

quasi-simple, 7
quaternion group, 8

self-centralising, 8
signalizer functor, 30
soluble signalizer functor, 30
special primitive pair of characteristic q, 13
symmetric group, 8

The F*-Structure Theorem, 91
The Soluble Z*-Theorem, 4
The Z*-Theorem for $\mathcal{K}_2$-groups, 5
Theorem A, 3
Theorem B, 4
Theorem C, 4
Theorem D, 5
type of a component, 8

unbalanced, 75

weakly balanced, 29

Editorial Information

To be published in the *Memoirs*, a paper must be correct, new, nontrivial, and significant. Further, it must be well written and of interest to a substantial number of mathematicians. Piecemeal results, such as an inconclusive step toward an unproved major theorem or a minor variation on a known result, are in general not acceptable for publication.

Papers appearing in *Memoirs* are generally at least 80 and not more than 200 published pages in length. Papers less than 80 or more than 200 published pages require the approval of the Managing Editor of the Transactions/Memoirs Editorial Board. Published pages are the same size as those generated in the style files provided for $\mathcal{A}_\mathcal{M}\mathcal{S}$-LaTeX or $\mathcal{A}_\mathcal{M}\mathcal{S}$-TeX.

Information on the backlog for this journal can be found on the AMS website starting from http://www.ams.org/memo.

A Consent to Publish is required before we can begin processing your paper. After a paper is accepted for publication, the Providence office will send a Consent to Publish and Copyright Agreement to all authors of the paper. By submitting a paper to the *Memoirs*, authors certify that the results have not been submitted to nor are they under consideration for publication by another journal, conference proceedings, or similar publication.

Information for Authors

Memoirs is an author-prepared publication. Once formatted for print and on-line publication, articles will be published as is with the addition of AMS-prepared frontmatter and backmatter. Articles are not copyedited; however, confirmation copy will be sent to the authors.

Initial submission. The AMS uses Centralized Manuscript Processing for initial submissions. Authors should submit a PDF file using the Initial Manuscript Submission form found at www.ams.org/submission/memo, or send one copy of the manuscript to the following address: Centralized Manuscript Processing, MEMOIRS OF THE AMS, 201 Charles Street, Providence, RI 02904-2294 USA. If a paper copy is being forwarded to the AMS, indicate that it is for *Memoirs* and include the name of the corresponding author, contact information such as email address or mailing address, and the name of an appropriate Editor to review the paper (see the list of Editors below).

The paper must contain a *descriptive title* and an *abstract* that summarizes the article in language suitable for workers in the general field (algebra, analysis, etc.). The *descriptive title* should be short, but informative; useless or vague phrases such as "some remarks about" or "concerning" should be avoided. The *abstract* should be at least one complete sentence, and at most 300 words. Included with the footnotes to the paper should be the 2010 *Mathematics Subject Classification* representing the primary and secondary subjects of the article. The classifications are accessible from www.ams.org/msc/. The Mathematics Subject Classification footnote may be followed by a list of *key words and phrases* describing the subject matter of the article and taken from it. Journal abbreviations used in bibliographies are listed in the latest *Mathematical Reviews* annual index. The series abbreviations are also accessible from www.ams.org/msnhtml/serials.pdf. To help in preparing and verifying references, the AMS offers MR Lookup, a Reference Tool for Linking, at www.ams.org/mrlookup/.

Electronically prepared manuscripts. The AMS encourages electronically prepared manuscripts, with a strong preference for $\mathcal{A}_\mathcal{M}\mathcal{S}$-LaTeX. To this end, the Society has prepared $\mathcal{A}_\mathcal{M}\mathcal{S}$-LaTeX author packages for each AMS publication. Author packages include instructions for preparing electronic manuscripts, samples, and a style file that generates the particular design specifications of that publication series. Though $\mathcal{A}_\mathcal{M}\mathcal{S}$-LaTeX is the highly preferred format of TeX, author packages are also available in $\mathcal{A}_\mathcal{M}\mathcal{S}$-TeX.

Authors may retrieve an author package for *Memoirs of the AMS* from www.ams.org/journals/memo/memoauthorpac.html or via FTP to ftp.ams.org (login as anonymous, enter your complete email address as password, and type cd pub/author-info). The

AMS Author Handbook and the *Instruction Manual* are available in PDF format from the author package link. The author package can also be obtained free of charge by sending email to `tech-support@ams.org` or from the Publication Division, American Mathematical Society, 201 Charles St., Providence, RI 02904-2294, USA. When requesting an author package, please specify $\mathcal{AMS}$-LaTeX or $\mathcal{AMS}$-TeX and the publication in which your paper will appear. Please be sure to include your complete mailing address.

After acceptance. The source files for the final version of the electronic manuscript should be sent to the Providence office immediately after the paper has been accepted for publication. The author should also submit a PDF of the final version of the paper to the editor, who will forward a copy to the Providence office.

Accepted electronically prepared files can be submitted via the web at `www.ams.org/submit-book-journal/`, sent via FTP, or sent on CD to the Electronic Prepress Department, American Mathematical Society, 201 Charles Street, Providence, RI 02904-2294 USA. TeX source files and graphic files can be transferred over the Internet by FTP to the Internet node `ftp.ams.org` (130.44.1.100). When sending a manuscript electronically via CD, please be sure to include a message indicating that the paper is for the *Memoirs*.

Electronic graphics. Comprehensive instructions on preparing graphics are available at `www.ams.org/authors/journals.html`. A few of the major requirements are given here.

Submit files for graphics as EPS (Encapsulated PostScript) files. This includes graphics originated via a graphics application as well as scanned photographs or other computer-generated images. If this is not possible, TIFF files are acceptable as long as they can be opened in Adobe Photoshop or Illustrator.

Authors using graphics packages for the creation of electronic art should also avoid the use of any lines thinner than 0.5 points in width. Many graphics packages allow the user to specify a "hairline" for a very thin line. Hairlines often look acceptable when proofed on a typical laser printer. However, when produced on a high-resolution laser imagesetter, hairlines become nearly invisible and will be lost entirely in the final printing process.

Screens should be set to values between 15% and 85%. Screens which fall outside of this range are too light or too dark to print correctly. Variations of screens within a graphic should be no less than 10%.

Inquiries. Any inquiries concerning a paper that has been accepted for publication should be sent to `memo-query@ams.org` or directly to the Electronic Prepress Department, American Mathematical Society, 201 Charles St., Providence, RI 02904-2294 USA.

Editors

This journal is designed particularly for long research papers, normally at least 80 pages in length, and groups of cognate papers in pure and applied mathematics. Papers intended for publication in the *Memoirs* should be addressed to one of the following editors. The AMS uses Centralized Manuscript Processing for initial submissions to AMS journals. Authors should follow instructions listed on the Initial Submission page found at www.ams.org/memo/memosubmit.html.

Algebra, to ALEXANDER KLESHCHEV, Department of Mathematics, University of Oregon, Eugene, OR 97403-1222; e-mail: klesh@uoregon.edu

Algebraic geometry, to DAN ABRAMOVICH, Department of Mathematics, Brown University, Box 1917, Providence, RI 02912; e-mail: amsedit@math.brown.edu

Algebraic topology, to SOREN GALATIUS, Department of Mathematics, Stanford University, Stanford, CA 94305 USA; e-mail: transactions@lists.stanford.edu

Arithmetic geometry, to TED CHINBURG, Department of Mathematics, University of Pennsylvania, Philadelphia, PA 19104-6395; e-mail: math-tams@math.upenn.edu

Automorphic forms, representation theory and combinatorics, to DANIEL BUMP, Department of Mathematics, Stanford University, Building 380, Sloan Hall, Stanford, California 94305; e-mail: bump@math.stanford.edu

Combinatorics, to JOHN R. STEMBRIDGE, Department of Mathematics, University of Michigan, Ann Arbor, Michigan 48109-1109; e-mail: JRS@umich.edu

Commutative and homological algebra, to LUCHEZAR L. AVRAMOV, Department of Mathematics, University of Nebraska, Lincoln, NE 68588-0130; e-mail: avramov@math.unl.edu

Differential geometry and global analysis, to CHRIS WOODWARD, Department of Mathematics, Rutgers University, 110 Frelinghuysen Road, Piscataway, NJ 08854; e-mail: ctw@math.rutgers.edu

Dynamical systems and ergodic theory and complex analysis, to YUNPING JIANG, Department of Mathematics, CUNY Queens College and Graduate Center, 65-30 Kissena Blvd., Flushing, NY 11367; e-mail: Yunping.Jiang@qc.cuny.edu

Functional analysis and operator algebras, to NATHANIEL BROWN, Department of Mathematics, 320 McAllister Building, Penn State University, University Park, PA 16802; e-mail: nbrown@math.psu.edu

Geometric analysis, to WILLIAM P. MINICOZZI II, Department of Mathematics, Johns Hopkins University, 3400 N. Charles St., Baltimore, MD 21218; e-mail: trans@math.jhu.edu

Geometric topology, to MARK FEIGHN, Math Department, Rutgers University, Newark, NJ 07102; e-mail: feighn@andromeda.rutgers.edu

Harmonic analysis, complex analysis, to MALABIKA PRAMANIK, Department of Mathematics, 1984 Mathematics Road, University of British Columbia, Vancouver, BC, Canada V6T 1Z2; e-mail: malabika@math.ubc.ca

Harmonic analysis, representation theory, and Lie theory, to E. P. VAN DEN BAN, Department of Mathematics, Utrecht University, P.O. Box 80 010, 3508 TA Utrecht, The Netherlands; e-mail: E.P.vandenBan@uu.nl

Logic, to ANTONIO MONTALBAN, Department of Mathematics, The University of California, Berkeley, Evans Hall #3840, Berkeley, California, CA 94720; e-mail: antonio@math.berkeley.edu

Number theory, to SHANKAR SEN, Department of Mathematics, 505 Malott Hall, Cornell University, Ithaca, NY 14853; e-mail: ss70@cornell.edu

Partial differential equations, to GUSTAVO PONCE, Department of Mathematics, South Hall, Room 6607, University of California, Santa Barbara, CA 93106; e-mail: ponce@math.ucsb.edu

Partial differential equations and functional analysis, to ALEXANDER KISELEV, Department of Mathematics, University of Wisconsin-Madison, 480 Lincoln Dr., Madison, WI 53706; e-mail: kisilev@math.wisc.edu

Probability and statistics, to PATRICK FITZSIMMONS, Department of Mathematics, University of California, San Diego, 9500 Gilman Drive, La Jolla, CA 92093-0112; e-mail: pfitzsim@math.ucsd.edu

Real analysis and partial differential equations, to WILHELM SCHLAG, Department of Mathematics, The University of Chicago, 5734 South University Avenue, Chicago, IL 60615; e-mail: schlag@math.uchicago.edu

All other communications to the editors, should be addressed to the Managing Editor, ALEJANDRO ADEM, Department of Mathematics, The University of British Columbia, Room 121, 1984 Mathematics Road, Vancouver, B.C., Canada V6T 1Z2; e-mail: adem@math.ubc.ca

Selected Published Titles in This Series

1055 **A. Knightly and C. Li**, Kuznetsov's Trace Formula and the Hecke Eigenvalues of Maass Forms, 2013

1054 **Kening Lu, Qiudong Wang, and Lai-Sang Young**, Strange Attractors for Periodically Forced Parabolic Equations, 2013

1053 **Alexander M. Blokh, Robbert J. Fokkink, John C. Mayer, Lex G. Oversteegen, and E. D. Tymchatyn**, Fixed Point Theorems for Plane Continua with Applications, 2013

1052 **J.-B. Bru and W. de Siqueira Pedra**, Non-cooperative Equilibria of Fermi Systems with Long Range Interactions, 2013

1051 **Ariel Barton**, Elliptic Partial Differential Equations with Almost-Real Coefficients, 2013

1050 **Thomas Lam, Luc Lapointe, Jennifer Morse, and Mark Shimozono**, The Poset of k-Shapes and Branching Rules for k-Schur Functions, 2013

1049 **David I. Stewart**, The Reductive Subgroups of F_4, 2013

1048 **Andrzej Nagórko**, Characterization and Topological Rigidity of Nöbeling Manifolds, 2013

1047 **Joachim Krieger and Jacob Sterbenz**, Global Regularity for the Yang-Mills Equations on High Dimensional Minkowski Space, 2013

1046 **Keith A. Kearnes and Emil W. Kiss**, The Shape of Congruence Lattices, 2013

1045 **David Cox, Andrew R. Kustin, Claudia Polini, and Bernd Ulrich**, A Study of Singularities on Rational Curves Via Syzygies, 2013

1044 **Steven N. Evans, David Steinsaltz, and Kenneth W. Wachter**, A Mutation-Selection Model with Recombination for General Genotypes, 2013

1043 **A. V. Sobolev**, Pseudo-Differential Operators with Discontinuous Symbols: Widom's Conjecture, 2013

1042 **Paul Mezo**, Character Identities in the Twisted Endoscopy of Real Reductive Groups, 2013

1041 **Verena Bögelein, Frank Duzaar, and Giuseppe Mingione**, The Regularity of General Parabolic Systems with Degenerate Diffusion, 2013

1040 **Weinan E and Jianfeng Lu**, The Kohn-Sham Equation for Deformed Crystals, 2013

1039 **Paolo Albano and Antonio Bove**, Wave Front Set of Solutions to Sums of Squares of Vector Fields, 2013

1038 **Dominique Lecomte**, Potential Wadge Classes, 2013

1037 **Jung-Chao Ban, Wen-Guei Hu, Song-Sun Lin, and Yin-Heng Lin**, Zeta Functions for Two-Dimensional Shifts of Finite Type, 2013

1036 **Matthias Lesch, Henri Moscovici, and Markus J. Pflaum**, Connes-Chern Character for Manifolds with Boundary and Eta Cochains, 2012

1035 **Igor Burban and Bernd Kreussler**, Vector Bundles on Degenerations of Elliptic Curves and Yang-Baxter Equations, 2012

1034 **Alexander Kleshchev and Vladimir Shchigolev**, Modular Branching Rules for Projective Representations of Symmetric Groups and Lowering Operators for the Supergroup $Q(n)$, 2012

1033 **Daniel Allcock**, The Reflective Lorentzian Lattices of Rank 3, 2012

1032 **John C. Baez, Aristide Baratin, Laurent Freidel, and Derek K. Wise**, Infinite-Dimensional Representations of 2-Groups, 2012

1031 **Idrisse Khemar**, Elliptic Integrable Systems: A Comprehensive Geometric Interpretation, 2012

1030 **Ernst Heintze and Christian Groß**, Finite Order Automorphisms and Real Forms of Affine Kac-Moody Algebras in the Smooth and Algebraic Category, 2012

For a complete list of titles in this series, visit the
AMS Bookstore at **www.ams.org/bookstore/memoseries/**.